THE **COMPLETE**
IDIOT'S
GUIDE TO

Android App
Development

by Christopher Froehlich

ALPHA

A member of Penguin Group (USA) Inc.

ALPHA BOOKS

Published by the Penguin Group

Penguin Group (USA) Inc., 375 Hudson Street, New York, New York 10014, USA

Penguin Group (Canada), 90 Eglinton Avenue East, Suite 700, Toronto, Ontario M4P 2Y3, Canada (a division of Pearson Penguin Canada Inc.)

Penguin Books Ltd., 80 Strand, London WC2R 0RL, England

Penguin Ireland, 25 St. Stephen's Green, Dublin 2, Ireland (a division of Penguin Books Ltd.)

Penguin Group (Australia), 250 Camberwell Road, Camberwell, Victoria 3124, Australia (a division of Pearson Australia Group Pty. Ltd.)

Penguin Books India Pvt. Ltd., 11 Community Centre, Panchsheel Park, New Delhi—110 017, India

Penguin Group (NZ), 67 Apollo Drive, Rosedale, North Shore, Auckland 1311, New Zealand (a division of Pearson New Zealand Ltd.)

Penguin Books (South Africa) (Pty.) Ltd., 24 Sturdee Avenue, Rosebank, Johannesburg 2196, South Africa

Penguin Books Ltd., Registered Offices: 80 Strand, London WC2R 0RL, England

Copyright © 2011 by Christopher Froehlich

International Standard Book Number: 978-1-61564-1-062
Library of Congress Catalog Card Number: 2010919337

13   14   15   8   7   6   5   4   3

Interpretation of the printing code: The rightmost number of the first series of numbers is the year of the book's printing; the rightmost number of the second series of numbers is the number of the book's printing. For example, a printing code of 11-1 shows that the first printing occurred in 2011.

Printed in the United States of America

Note: This publication contains the opinions and ideas of its author. It is intended to provide helpful and informative material on the subject matter covered. It is sold with the understanding that the author and publisher are not engaged in rendering professional services in the book. If the reader requires personal assistance or advice, a competent professional should be consulted.

The author and publisher specifically disclaim any responsibility for any liability, loss, or risk, personal or otherwise, which is incurred as a consequence, directly or indirectly, of the use and application of any of the contents of this book.

Most Alpha books are available at special quantity discounts for bulk purchases for sales promotions, premiums, fund-raising, or educational use. Special books, or book excerpts, can also be created to fit specific needs.

For details, write: Special Markets, Alpha Books, 375 Hudson Street, New York, NY 10014.

Publisher: *Marie Butler-Knight*
Associate Publisher/Acquiring Editor: *Mike Sanders*
Executive Managing Editor: *Billy Fields*
Development Editor: *Ginny Bess Munroe*
Senior Production Editor: *Janette Lynn*
Copy Editor: *Daron Thayer*

Cover Designer: *Rebecca Batchelor*
Book Designers: *William Thomas, Rebecca Batchelor*
Indexer: *Heather McNeill*
Layout: *Ayanna Lacey*
Senior Proofreader: *Laura Caddell*

Contents

Part 1: Getting Started 1

1   An Open Invitation 3
Starting from Scratch 3
    Software 4
    Programming Experience 4
Selecting Your Development Environment.................5
Varieties of Devices...6
Selecting a Platform 7
    Device Restrictions..8
    Forward Thinking and Backwards Compatibility.............9
Assembling Your Toolkit 9
    Java..10
    Eclipse IDE for Java Developers 10
    Android SDK 11
    The ADT Plugin for Eclipse 12
    Registering for the Android Market.........................14

2   Building for Android 15
Android Development Overview......................................15
    Introduction to Eclipse...16
    The Role of the Android SDK 18
Starting Your First Project 18
    Creating the Project Files...19
    Inside Your Project..21
    Setting Your Preferences 22
Aloha, Android 24
    Meet the Emulator 25
    Debugging with the Emulator 27
Making the Device Connection......................................28

3   Crafting the Layout 31
Application Fundamentals 32
Reintroducing the Recipe App 34
    Laying Out the Welcome Screen..................................34
    Expanding Your View...40
    Formatting for All Displays...46
Your Icon...47
Introducing Your First Error 49

4   Finding More Activities **53**

Improving Your Recipe App 53
   Introduction to Refactoring *54*
   About Your App *56*
   Updating the Android Manifest *57*
   Adding an Activity *58*
   Creating the Layout *59*
   Putting It Together *62*
Styling Your App 65
   Selecting a Theme *65*
   Rolling Your Own Style *67*
Polishing Your App 68
   Adding a Menu *68*
   Storing Some Settings *71*
   Quitting Time *75*

5   Programming for Android **77**

What Is Java? 77
   Object Oriented Programming Review *78*
   The Java Virtual Machine *79*
The Front of an Android Application 80
   Activities *80*
   Views *82*
Behind the Scenes 86
   Intents *86*
   Intent Filters *88*
   Background Receivers *89*

Part 2:   Constructing Your Application **91**

6   Resources and Animation **93**

The Splash View 93
   Assemble the View *94*
   Colors and Dimensions *96*
   Ready, Set, Animate *99*
   Method Acting *101*
2D Graphics Overview 106
   Working with Drawables *107*
   Using Shapes *107*

7 Building Input and Output 111

Making a New Recipe 111

 EditText and Other Widgets 112

 Deep Nested Views 115

Working with ListViews 120

 Inflating the Layout 120

 From XML to Java 123

Simplify the Interface with TabHosts 127

Toast Feedback 131

8 Storing and Retrieving Data 135

Storage Overview 135

 SharedPreferences 136

 Internal Storage 138

 External Data 140

Saving a New Recipe 140

 Retrieving SharedPreferences 141

 Storing Input as SharedPreferences 148

Manage Your Recipes 151

9 Search for It 157

Search Basics 157

 Create the Configuration 159

 Define a Searchable Activity 161

 Build the Search Activity 162

 Providing the Search Option 165

Voice Search 166

More Search Options 167

10 From Widgets to the Browser 169

Widgets and Dialogs 169

 Constructing a Dialog 171

 Dates and Times 173

 More Widgets and Controls 176

Integrating the Browser 179

 Opening the Web Page 180

 Linkify 182

 Embedding the Browser 183

Part 3:   Make the Most of the Hardware 189

11   Cameras and Media..**191**
Introduction to Multimedia on Android 191
Incorporating the Camera into Your App 193
  Prepare the SurfaceView...*193*
  Working the Camera Class *195*
  Capturing the Photo...*197*
  Recording Video *198*
Accessing Stored Photos 200
Embedding Audio or Video...201
Using the Microphone...202

12   Location-Based Services **205**
Introduction to Android Sensors and Receivers 205
Location, Location...207
  Location Services..*208*
  Consider Accuracy and the Battery...................................*212*
  Emulator Limitations *213*
Sense and Sensors...214
  Reading a Sensor *215*
  Sensor Data..*218*
Integrate Maps..219
  Register for Google Maps Access *219*
  Working with Google Maps...*220*

13   3D Graphics and Animation.................................**225**
Introduction to 3D Graphics...225
  OpenGL Basics *226*
  Drawing an Activity..*228*
Model Objects 230
  The Cube *230*
  Moving the Cube in 3D *233*

14   Core Services..**239**
Overview of Android Hardware...239
The Phone Itself..240
  Working with Calls..*241*
  Send SMS and MMS Messages..*245*

Network Interfaces 247

   Wi-Fi vs. 3G *248*

   Wi-Fi *250*

   Working with Bluetooth *251*

Part 4: **Increasing Your Application Scope** **259**

15 **A Touch of Locale** **261**

   Localization 261

     Managing Resources *262*

     Alternate Resources *264*

     Performing Translation in Code *267*

   Runtime Changes 270

     Efficiently Rotating the Screen *272*

     Manually Manage Configuration Changes *273*

   Multi-Touch 274

     MotionEvents *275*

     Build Your Own Gestures *277*

16 **Threads and the Background** **283**

   What Are Threads? 283

     About Processes *285*

     Working with AsyncTask *286*

     Dispatching Threads *290*

   Services 293

     The Service Life Cycle *294*

     Building a Basic Service *294*

   Using Notifications 297

17 **SQL and Databases** **301**

   SQL at a Glance 301

   Intro to SQLite 304

     Building the Helper Class *305*

     Database Interaction *307*

     More SQL Detail *310*

     Bring the Data into View *311*

   Additional SQLite Tools 315

18   ContentProviders...**319**

Introduction to ContentProviders 319

Provider Syntax.. *320*

Requesting Data from a ContentProvider *322*

Building Your Own Provider 323

Prepare the Class *324*

Implement the Methods... *326*

Provider Permissions and the Manifest *330*

Part 5:   Taking Your App to Market...............................**333**

19   Comprehensive Debugging **335**

Logging...335

Debugging Your App 338

The Debug Perspective.. *339*

Advanced Logging with Traceview *340*

Switching to the DDMS Perspective *341*

Debugging with Dev Tools....................................... *343*

20   Testing Your Apps **345**

The Good News... 345

Build a Testing Plan.. 346

Selecting Your Targets *347*

Working with Multiple Versions of Android..................... *348*

Methods Change... *350*

Be Mindful of the Future *352*

And More Tests 353

21   App Markets and Beyond **355**

Available Android Markets 355

Welcome to the Android Market 357

Registering Your Accounts....................................... *357*

App Preparation *358*

Certification *360*

And Release! 362

Next Steps.. 364

Advertising.. *365*

The Next Generation *366*

Keep in Touch *366*

Appendixes

A   Glossary 369

B   Resources, References, and Useful Websites....375

      Index 379

Introduction

Learning to develop software is like learning to ride a bike: you will not get very far unless you hop on and start pedaling. This book focuses on development in much the same way. Each chapter will introduce new, core concepts to Android and follow with code samples. I find that I learn best by sitting down and writing code. I encourage you to follow this book with your computer, writing and testing code as you go. Please view this book more as a guide to follow than a textbook to memorize.

The first two parts of the book will walk you through developing and extending a single, core app. Each chapter in these parts is designed to continue building on the one before. The final three parts compartmentalize the topics, providing functional sample code that can be run independently. All chapters are brief and designed to focus on the most important content in-depth while still providing contextual information for the whole topic.

Developing Android apps is an exciting space to occupy. New Android devices and features are developed rapidly, providing you with new platforms and capabilities to implement your ideas. You may or may not already be familiar with Java and the Eclipse development environment, but you will likely enjoy meeting the new challenges that Android has to offer. From the nuances of the development kit and Android's unique operating system mechanics to publishing your app in the Market, you have a wealth of fun opportunities to learn ahead of you.

Conventions Used in This Book

When you need to write code in Android, it takes one of two forms: XML or Java. With XML, all code samples in this book have the following syntax:

```xml
<?xml version="1.0" encoding="utf-8"?>
<menu
  xmlns:android="http://schemas.android.com/apk/res/android">
    <item
        android:id="@+id/main_menu_new"
        >
    </item>
  </menu>
```

All Java code is formatted like this:

```java
public FabulousNewClass( String message ) {
log.d( TAG, message );
  }
```

The steps of instructions are formatted in **bold.** For example, to create a new Android project in Eclipse, select **File > New > Android Project** from the menu.

How This Book is Organized

This book is divided into five parts. The first few chapters introduce Android and its capabilities, the preferred developer tools, and the Android SDK, which you use as the basis for your app development. Once you have your first app built and running, the chapters introduce laying out your app, working with resources, and building user interfaces. As you become comfortable working with screens and widgets, you learn how to incorporate core hardware features like the camera and location services. After you have explored localization and databases, you will be ready to prepare your app for the Android Market.

Part 1: Getting Started. Here you download the Android SDK and developer tools to begin building your apps. Get acquainted with Eclipse and the Android emulator, tools you use in every step of the process. You write your first app and explore Java and the Android SDK.

Part 2: Constructing Your Application. Accelerate the development of your user interfaces with animations and dialogs. You learn how to store and retrieve information, and how to provide search capability. Fundamental concepts in user interaction are covered in this part.

Part 3: Make the Most of the Hardware. Android devices are chock full of powerful hardware features. From location services to Bluetooth and cameras to 3D animation, here you learn how to allow your app to make the most of its physical device.

Part 4: Increasing Your Application Scope. One of your goals as an app developer is probably to reach as large an audience as possible. This part demonstrates how to localize your app for different languages and regions, store data more efficiently, and how to enable your app to communicate with other Android apps.

Part 5: Taking Your App to Market. If you have started app development, you likely want to be able to distribute your app. The Android Market is the largest Android app storefront in the world, and this part takes you through the process of preparing your app for Market. By fixing errors through debugging and testing your app, you will be ready to begin the process of submitting your app for public consumption.

Extras

Throughout the book you will see sidebars that elaborate on keywords, principles, and ideas.

DEFINITION

Key terms in Android app development.

PITFALL

Tips to avoid common mistakes.

ANDROID DOES

Advice to save you time and effort.

GOOGLE IT

Interesting factoids to encourage your development growth.

Acknowledgments

I would like to thank my wife Heather and my son Eli, who have gone above and beyond in supporting me through the adventure of writing this book. Thank you both for your encouragement.

Additionally, thank you David, Phil, and Steve for your support and guidance.

Special Thanks to the Technical Reviewer

The Complete Idiot's Guide to Android App Development was reviewed by an expert who double-checked the accuracy of what you'll learn here, to help us ensure that this book gives you everything you need to know about Android development. Special thanks to Cliff Lardin and Damon Brown.

Trademarks

All terms mentioned in this book that are known to be or are suspected of being trademarks or service marks have been appropriately capitalized. Alpha Books and Penguin Group (USA) Inc. cannot attest to the accuracy of this information. Use of a term in this book should not be regarded as affecting the validity of any trademark or service mark.

Getting Started

How do you get started writing apps for Android? What software do you need? What programming languages does Android support? How do you get started with Eclipse?

This part answers these questions and begins your transition into Android development. By the end of this part, you will be familiar with Eclipse and creating your first Android apps.

An Open Invitation

In This Chapter

- What you need to get started
- Android features and limitations
- The Android Software Development Kit (SDK)

As a software developer, I spend a lot of time with other developers brainstorming application ideas. Imagine a laser tag application that runs on your mobile phone, uses GPS coordinates to locate you and all of your friends, then layers their profile pictures over a street view map of your area. Using a digital compass and accelerometers, you can play virtual laser tag all over your home town. Neither the software nor the hardware existed years ago when I first imagined this type of app, but Google's Android OS has changed many of our perceptions about what is possible.

Your imagination is very nearly the only limit to what types of apps you can develop. Android is now the cutting edge for mobile app developers, on phones, tablets, and Internet connected TVs. As a developer, I could not be more excited about the opportunities Android has to offer. If you feel the same way, I will be here to assist you with the process of developing your apps through publication to the Android Market. Before diving in, there are a few things to prepare in advance.

Starting from Scratch

While you are probably holding back a flood of great app ideas that you want to start translating into fully baked Android applications, there are a few things to observe before proceeding.

Software

There are very few hardware limitations to developing for Android, and you can use any modern operating system. In fact, as long as you can install and run *Eclipse*, you have met the core requirement to develop for Android. Additional tools are required, but they are free and easy to acquire. If you purchased your computer within the last five years, at the time of this writing, you are more than likely equipped with everything you need. If your computer is older, try Eclipse anyway; you may be in luck or might need to add only memory to your system.

> **DEFINITION**
>
> **Eclipse** is the preferred Integrated Development Environment (IDE) for Android development. An IDE is a toolkit for writing, debugging, compiling, and deploying applications.

Furthermore, you do not even need to own an actual Android phone. The *Android Software Development Kit (SDK)* comes with a device emulator that enables you to run your apps against different versions of the Android Operating System (OS). As you mature as a mobile developer, you will want to begin testing and deploying apps on physical hardware; but you can get started with Android right now.

> **DEFINITION**
>
> The **Android SDK** is a collection of tools that integrate into your development environment enabling you to develop applications for any device (phone, tablet, or netbook) which runs the Android OS.

Programming Experience

While it is difficult to pinpoint exactly which skills a developer needs to succeed with Android, you should be familiar with the following concepts and languages:

- **General programming knowledge.** You do not need a Ph.D. in software engineering to start, but you should be comfortable with concepts such as "functions," "variables," "conditional logic," and "references."

- **Some familiarity with C, C++, or Java.** Android apps must be compiled in Java, but experience with any of these languages will make your transition smooth.

- **A basic understanding of Object Oriented Programming (OOP).** You should grasp concepts such as "classes," "interfaces," "methods," and "objects." Chapter 5 covers some of these concepts if you need to refresh your memory.

ANDROID DOES

Check the resources in Appendix B to find some fantastic sites that can supplement your Android programming skill set.

Do not despair if you feel your skills are out dated or if you don't have experience with any of these requirements. Throughout the book, as you encounter complex and difficult topics, secondary resources are provided to expand upon the depth of the subject matter.

Selecting Your Development Environment

Unlike other mobile operating systems, Android OS is based on *Linux*, which means that the OS is open source and free to distribute and modify. Google's vision for the future includes Android running not just on mobile phones, but in your car and on your TV.

DEFINITION

Linux is a versatile, robust, and free operating system that is used to power a wide range of devices, from computers to TiVos and now mobile phones.

To ensure that the app you write today works on present and future Android powered devices, Google wisely chose Java to be the standard platform for native code. This frees you, the developer, to choose any development platform that can load the Java Runtime Environment and the Java SDK. Selecting your development environment is as simple as choosing the computer you like most, probably the one you already own.

While it is true that practically any modern computer and operating system will meet your development needs, it is not possible in this text to walk through the subtle differences in the Android SDK on all possible systems. Therefore, for the purpose of this book, we will use screenshots of the various tools taken from a Microsoft Windows system.

If you are a Mac or Linux user, fret not! The images on your screens will be nearly identical, and the links provided to download the tools also contain more detailed guidelines specific to your systems. To reduce confusion, this guide also assumes that you are familiar with the basics of installing applications on the systems of your choice.

GOOGLE IT

All of the software tools needed to develop for Android are documented on Google's own help pages. If you get stuck, Google it!

Varieties of Devices

Android powers over 40 different devices, each with different hardware features. With so many differences and choices, what makes an Android device, well, an Android device? Most devices share the following features:

- Multi-Touch Screen
- Internet
- Gmail, Contacts, Calendar
- Multimedia Support
- Sensors and Accelerometer
- Compass
- Location Awareness
- Camera
- Bluetooth
- 2D/3D Graphics
- Microphone

PITFALL

Not every device has all of these features, and some devices have more. For example, the HTC EVO has dual-facing cameras, whereas the Motorola Droid X has only one. You should anticipate users running your app with and without specific hardware components.

Just think of all the applications for these features! With multimedia support and Internet connectivity, you get apps like Pandora and Last.fm. With location awareness and a camera, you get Google Goggles, an app that will identify and even translate the text on anything you take a picture of. Sophisticated apps abound in the Android Market, providing turn-by-turn directions, real-time Twitter and Facebook feeds, advanced voice recognition services, and casual and advanced games. Your creativity is the only boundary for developing Android applications.

Selecting a Platform

If you have ever developed an application for any other platform, you will be familiar with the tension between the following desires: you want to use the latest and greatest features of the platform, but you also want your app to work on as many different devices as possible. Why do companies like Google force us to make these difficult decisions? The simple answer is, no one is perfect. For example, Google did not anticipate all of the ways developers would use multitasking as a feature when it was released, and changes to the Android SDK have required breaking functionality in order to repair and improve it.

As a developer, this means that you must invest some time and energy researching the current market. Ask yourself these questions:

- Do I need to fully support every version of Android?
- What are the most popular versions of the Android OS?
- How many Android OS versions do I need to support to succeed?

ANDROID DOES

When you begin testing your application, the Android SDK will allow you to compile your app against any or all versions of the OS, so you will always be able to test against specific Android releases.

Device Restrictions

Think more about the differences between what you might consider traditional programming, either desktop apps or web applications for desktop browsers. Modern computers come standard with larger and larger displays and faster, more powerful processors. Mobile devices improve constantly, but mobile development requires thinking about maximizing less powerful hardware inside smaller spaces.

The first Android device, the T-Mobile G1, had a 3.2-inch screen with a resolution of 320 × 480 pixels. The HTC EVO, one of the larger mobile phone screens, has a 4.3-inch screen with 800 × 480 pixels. From small to large, this is still not a huge amount of screen real estate to construct your app.

These types of constraints should inform and drive your creativity. Unlike a desktop, which is more likely to always have Internet connectivity, a mobile device is traveling through buildings, tunnels, across the country where Internet connectivity is probably intermittent. While even our laptops now come standard with hundreds of gigabytes of memory, mobile phones still expand only to 32 or 64 gigabytes. Begin to think about these limitations as opportunities to innovate.

Beyond these physical restraints, what other restrictions does Android impose on its developers? Following is a list.

- **True multitasking has a cost.** Your app can run background threads and even services, which run when the app has closed; but Android maintains a list of priority functions. If your background thread or process takes too long to complete or begins to consume too much memory, Android kills it.

- **Input frustrations.** Of the Android devices that implement hardware keyboards, the keyboards are often small and more difficult to use. Multi-touch support, although generally great for the latest versions of Android hardware, is not always well implemented on older devices.

- **Living in a sandbox.** All Android apps run as self-contained units with no access to other applications. The Android SDK provides a means of communicating, but it requires implementation that other developers may not have done.

- **Varied screen real estate.** The screens on Android mobile phones are small, tablets are larger, and TVs are huge. When you develop an Android app, you must consider the screen size variations.

- **Your apps will crash.** It is inevitable and frustrating. Android prioritizes certain apps, like the phone and SMS messaging; Android can and will make sure these apps take priority, even at the cost of your own app. Anticipating different use cases of your app is critical to success.

While these and other obstacles present road blocks to your development success, the Android SDK provides numerous resources to prepare for and protect from Android's built-in protections and the unpredictability of mobile device usage.

Forward Thinking and Backwards Compatibility

As you may have already begun to realize, one of Android's greatest strengths is its support for many different types of devices with different features. This is also one of the challenges for developing for Android; selecting any one version of the operating system for development increases the chances of compatibility issues for users wanting to run your app on their devices. For the purposes of this guide, we use Android OS 2.2. As you become comfortable developing and debugging your apps, you will find it easier to support multiple versions of Android simultaneously.

ANDROID DOES

The Android SDK provides emulators not only for the version of Android OS you currently target, but also all versions of Android OS. This allows you to quickly target different versions and thoroughly test your app.

Like them or leave them, these are the realities of Android development. Most developers and I agree that they are small obstacles to overcome and that you will quickly begin seeing many of them as strengths and opportunities.

Assembling Your Toolkit

We have dabbled in theory long enough. Let us start getting our hands dirty with some development work! There are quite a few different tools you will need to download and install in order to configure your development environment. Open your browser of choice, such as Firefox or Chrome, and let us begin.

GOOGLE IT

The single most important web resource for you as a developer is Google's own Android for developers site at http://developer.android.com. Bookmark and visit this site frequently.

Java

Before you can do any development work on Android, you must have the latest Java Runtime Environment (JRE) and the Java SDK (JDK). Both are available at http://www.oracle.com/technetwork/java/javase/downloads. Links can change, and if the link no longer works for you, Google it.

If you are a Windows or Linux user, you can download them directly from the website. However, if you use a Mac, it will direct you to Apple and have you register with the Apple website as a developer. Choose one of the free developer registration options and download the Java software.

After the downloads are complete, launch the installers for the JRE and JDK respectively and complete the installation steps.

Eclipse IDE for Java Developers

To continue, you will need an *IDE*. The most widely used and supported application is the Eclipse IDE for Java Developers 3.6 (Helios), which you can download at http://www.eclipse.org/downloads.

DEFINITION

An Integrated Development Environment or **IDE** is an application designed to assist your code writing effort by helping compile your application, identify errors, and cross-reference functionality. IDEs help improve development efficiency.

The download will complete as a zip file. Extract this to C:\eclipse or the location of your choice, then double-click the Eclipse application.

You will be prompted to select a default workspace, as shown in Figure 1.1.

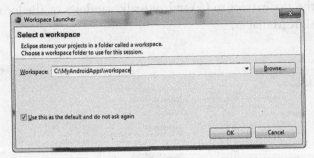

Figure 1.1: *Select a default workspace to save your program.*

Enter your preferred location or accept the default and continue.

Android SDK

With one more step to go, you are now ready to download and install the Android SDK. Navigate to http://developer.android.com/sdk, as shown in Figure 1.2.

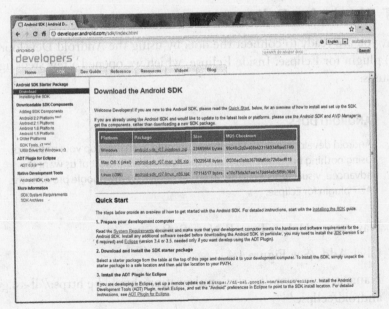

Figure 1.2: *The Android developer website.*

Because the Android SDK supports all versions of the Android OS, you will only see one download option for your operating system (see Figure 1.3). Once you have downloaded the zip, extract to C:\android-sdk-windows or the location of your choice and execute SDK Manager.

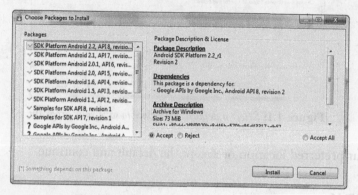

Figure 1.3: *Download the Android SDK.*

Accept the default selections, and then start the installation process.

The ADT Plugin for Eclipse

Finally, you are ready to connect the dots by using the Android Development Tools (ADT) plugin for Eclipse. Inside Eclipse, which we opened a few steps ago, complete these steps:

ANDROID DOES

Android development is so flexible, you could even develop your entire app using nothing but a text editor and the Android SDK. Most of us want the advanced, visual tools available in an IDE, which is why Google provides the ADT plugin for Eclipse.

1. Navigate to **Help > Install New Software.**

2. Click **Add.** (See Figure 1.4.)

3. Name is your own description. Location should be https://dl-ssl.google.com/android/eclipse/.

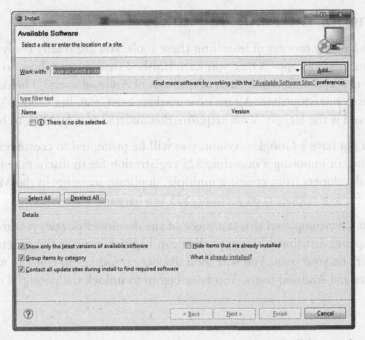

Figure 1.4: *Adding in the Android Development Tools to the Eclipse workspace.*

4. Select **Developer Tools > Next** and wait for the software to connect to the Android website.

    Confirm the selected options.

    Agree to the license terms.

5. Eclipse prompts you to restart.

6. After Eclipse opens again, navigate to **Windows > Preferences** (on Windows) or **Eclipse > Preferences** (on Mac).

7. From the left, select **Android.**

8. You will receive a pop-up window asking to send usage statistics to Google. Your response is optional.

9. Browse for or enter the Android SDK location, and click **Apply.**

10. Select **Android 2.2** and click **OK.**

Registering for the Android Market

After the lengthy process of installing these tools, you are finally ready to begin developing Android apps. Once you have finished that journey, how do you move your hot new app onto the devices of millions of Android users? The Android Market is your first, one-stop shop. Alternative markets exist, too, but Google's Market was the first and is the largest. Visit http://market.android.com/publish to begin.

If you do not have a Google account, you will be prompted to create one. Google recently began imposing a one-time $25 registration fee to discourage less than altruistic app developers from creating multiple, duplicate accounts in the Market. While not free, for a developer class account, $25 is a bargain.

Once you have completed this last piece of the developer puzzle, you are at last free to develop and distribute your apps. You can begin to explore the functionality of the Market on your own. You will soon discover resources to connect with other developers and Android users. You have begun to unlock the potential of the Android platform.

The Least You Need to Know

- You can develop for Android with any operating system which can run Eclipse.
- The Android platform presents an opportunity to think about application development in a very different way.
- Remember and reference with fervent frequency the Android Developer Site at http://developer.android.com.
- All of the developer tools for Android are free.
- Registration for an Android Market account costs $25.

Building for Android

In This Chapter

- Getting to know Eclipse
- Creating projects in Eclipse
- Running apps in the Android emulator
- Starting your first project

If you have not developed an app for another flavor of mobile phone, you have probably installed enough apps on your existing mobile devices to know the potential that mobile development offers. From card games to jogging assistants and from news readers to real-time astronomy guides, every type of mobile app imaginable is either waiting for you to download or create it.

Finding and using apps is usually straightforward and intuitive, but how do you translate your app ideas into reality? Most of your life as a mobile app creator is spent developing your application, from inception to delivery. All of this work happens inside your development environment, which will soon become your greatest ally. In this chapter, you explore Eclipse, create your first Android app, and run the Android emulator to experience the complete development life cycle.

Android Development Overview

The first step to beginning any task is to collect the necessary tools. In Chapter 1, you downloaded and installed the Eclipse IDE, the Android SDK, and the ADT Eclipse plugin. While powerful and complex, each of these tools serves a simple, basic purpose. At its core, Eclipse is simply an editor, like an advanced word processor, enabling you to write your app. The Android SDK is the toolkit for communicating in the native Android language, and the ADT plugin is a mechanism to automate the translation of the code you write in Eclipse into the language of the Android OS.

Introduction to Eclipse

Eclipse was created to assist Java developers, and it has gained popularity over the years for its flexibility and ease of use. If you have ever used Visual Studio on Windows, XCode on Mac, or any other development environment, Eclipse will quickly become familiar to you.

As you noticed from Chapter 1, Eclipse does not install like other programs. To launch it, open your /eclipse directory and launch **eclipse.exe** or double-click the Eclipse icon in the visual environment. You may want to create a Desktop shortcut to the program for quick access later.

As shown in Figure 2.1, the Welcome screen displays some orientation information about Eclipse, including tutorials and sample code. This screen displays only the first time you open Eclipse. If you need to come back to it later, **Help > Welcome** will bring you back. Let's open your Workbench to get started.

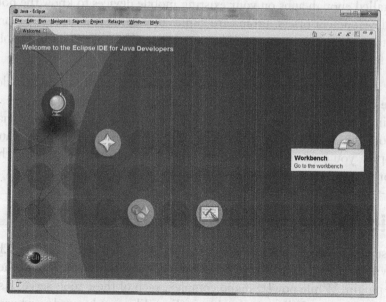

Figure 2.1: *The Eclipse Welcome screen.*

GOOGLE IT

The Eclipse Foundation, maintainers of the Eclipse project, provides extensive tutorials and guides for beginners to advanced users. Visit http://eclipsetutorial. sourceforge.net to explore all the possibilities that Eclipse can offer.

You will soon be creating projects and filling this space with content, but take a moment to orient yourself. You should note that all of the mini-windows or *views* inside your Workbench are independently configurable. You have complete control over this workspace. See Figure 2.2.

DEFINITION

Eclipse uses **views** to display different types of information about your Workbench. Each view has its own set of preferences and menu options and can be customized to your needs.

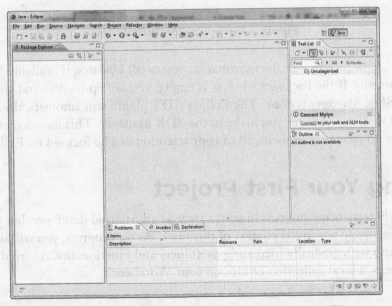

Figure 2.2: *The Eclipse workspace, where your projects will be.*

The sum of all the views in your current Workbench is referred to as a *perspective*. Eclipse provides a few default perspectives to work with, and from the Eclipse Window menu you can save the changes you make to your views as new perspectives for persistence.

The Role of the Android SDK

As you may have noticed, the Android SDK is still a separate product. Generally, you will only need to interact with it outside of Eclipse to download updates from Google. You can launch it by executing **SDK Manager.exe,** probably in your /android-sdk-windows directory (see Figure 2.3).

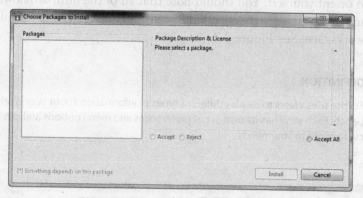

Figure 2.3: *Installing the Android SDK.*

This window provides all the information you need. Updates, if available, will appear automatically. If the Packages window is empty, you are up-to-date and your work in the SDK Manager is done. The Eclipse ADT plugin will automatically refresh Eclipse with any changes you make in the SDK Manager. This means that once you have set up your environment, all of your attention can be focused on Eclipse.

Starting Your First Project

You will create many projects in your career as an Android developer, but you will create relatively few in the context of this book. As we progress, you will develop a recipe app with gradually increasing usefulness and functionality. Android apps begin as projects, a local collection of files on your Workbench.

 GOOGLE IT

The source code and project files for all of the code content in this book are freely downloadable from http://code.google.com/a/eclipselabs.org/p/cig-android-development where you can download files by chapter or collect the whole source.

Creating the Project Files

From the main menu in Eclipse, select **File > New > Project > Android > Android Project**.

The **New Android Project** window offers a number of configuration possibilities (see Figure 2.4).

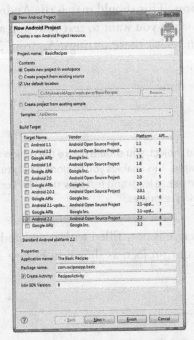

Figure 2.4: *Starting a new Android project.*

Here are the parameters:

- **Project name.** This should be a descriptive name for the entire project. For our test run, we'll call it "BasicRecipes".

- **Build target.** Which version of the Android OS do you want to write your app for? This depends on your market research, which Android features you must have, and what devices you want to target. For this project, Android 2.2 will work.

- **Application name.** This should be the name of the actual application, as you want it to appear in the title bar.

- **Package name.** This refers to the Java namespace which this project will implement. Your projects here will use the com.recipesapp namespace.

- **Create Activity.** This instructs Eclipse to create a default activity for you. Activities are one of the core building blocks of an Android application and are discussed in greater depth in Chapter 3. We can just type in "RecipeActivity" for now.

- **Min SDK Version.** This should match the Android version you have selected to target. Later, when you want to test your app against different versions of Android, you can revisit these settings. Android 2.2 is SDK Version 8.

Click **Finish** and you are done.

ANDROID DOES

In the New Android Project, you should notice the option to **Create project from existing sample.** This enables you to clone a fully functional example app, which you can use to learn and test new skills. Try LunarLander to get an early peek at drawing resources.

Your Workbench perspective is now filled with populated views containing the most relevant information to your project, as shown in Figure 2.5.

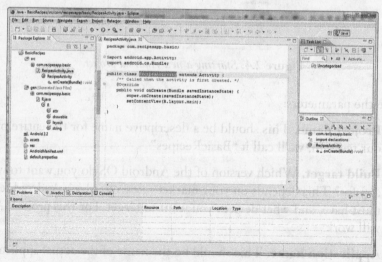

Figure 2.5: *The various views within your Workbench.*

Here are the elements:

- **Package Explorer view.** Displays the Java elements in your open projects. You can have multiple projects open at the same time; they all appear in the same view. Most of this content is internal and comes directly from your workspace folder, but some is external and will reference libraries from other sources. Notice the Android 2.2 folder derives from the Android SDK folder. Think of the Package Explorer as the collection of all libraries related to your project.

- **Editor view.** Displays the content of the code you open from Package Explorer. This view features syntax highlighting, code assistance, and commenting. All of the code you write in Eclipse will occur in the Editor view.

- **Tasks List view.** Shows tasks which you have created and assigned to your project. Use Tasks to organize the road map of your app development.

- **Outline view.** Renders the contents of the currently selected Editor view as a tree, organizing your source code according to classes, fields, and methods.

- **Problems view.** Displays collections errors, warnings, and problems that arise out of your code as you build. If your code contains syntax errors or invalid calls, the Problems view logs and organizes these errors by severity.

PITFALL

If you receive an error in Eclipse in the Problems tab after creating a new project, such as "Project 'BasicRecipes' is missing required source folder: 'gen'", have no fear. This is a common Eclipse bug. Closing and reopening Eclipse will correct the issue.

Inside Your Project

You can view your project in two ways, from your Workbench folder on your file system, or inside Eclipse under Package Explorer as you discovered previously. First, take a look at the key files and folders Eclipse has created for your project.

Important Android Project Files

Folder/File	Description
/	The root folder of your Android project, in this case /BasicRecipes.
/AndroidManifest.xml	This *XML* file defines the components of your Android app—activities, services, intents—as well as permissions and capabilities.
/assets	This folder manages references to static files you wish to include in your app.
/bin	A folder including a built-up compiled version of your app.
/gen	A compilation of your project's resources which become accessible using the R class.
/res	Your resources folder, where your animation, layout, menu, and XML data will be collected.
/src	The source code folder for your app. This will contain the code that you write.

DEFINITION

Extensible Markup Language (XML) is a web standard for storing data of any kind in a format easy to process programmatically. Almost all Android configuration is done with XML formatted documents.

Notice that the Package Explorer tree differs slightly from the folder tree on disk. Eclipse hides the /bin folder from view, and the Android 2.2 appears, though it does not exist on disk. This folder is linked into your project automatically on creation to give you immediate access to all of Android's APIs.

Setting Your Preferences

You have yet to begin writing your app, but you can save yourself some time and frustration by making a few adjustments to the default settings in Eclipse. These Eclipse settings are the first you will likely want to change, which are highlighted in Figure 2.6.

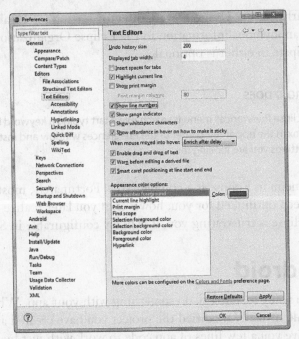

Figure 2.6: *The Preferences screen that allows you to set up your environment.*

All of your Eclipse environment settings are accessible from the **Window > Preferences** menu (on Windows) or the **Window > Preferences** (on Mac). From this settings view, you will likely want to customize these settings. Follow these steps:

1. Choose **General > Editors > Text Editors** and check **Show Line Numbers.** This setting is often off by default, but you will find it useful when debugging your code to see the line number referenced in your console output. You will find that whether working with the debugger or other developers, line number references are frequent and helpful to have quick access to.

2. Still in **General,** choose **Appearance > Colors and Fonts.** Expand **Basic** to change the visual style for all Eclipse code or just edit **Java** to limit your changes to Android development. You will spend a considerable amount of time looking at color-coded, formatted text; do yourself an early favor and make sure these colors are easy on your eyes.

3. In **Android > Usage Stats,** check or uncheck **Send usage statistics to Google.** This determines whether to send anonymous information about your Android development to Google for their use in improving the product.

Similarly, set the **Usage Data Collector** setting for **Enable capture,** which sends anonymous information to the Eclipse Organization. Your choice to participate in either is optional.

ANDROID DOES

The Eclipse Preferences menu is searchable. Start typing a keyword into the text box in the upper-left corner of the Preferences window and instantly find the settings you are looking for.

The Settings menu in Eclipse is a complex tree. Fortunately, most of these settings have already been optimized for you; however, if you later realize that a particular behavior in Eclipse is frustrating you, it is likely configurable in Settings.

Aloha, Android

At last, you are ready to actually do something with your app. While you have not written any code or really modified the project you have created and explored, Eclipse does give you a few lines of app code to work with just by creating a project. Eclipse generates *Hello World* for you when you elect to create a default **Activity, RecipeActivity** from previously. The code is located in the main.xml file.

DEFINITION

Hello World is often the first program developers write for new platforms or languages. The program is intended to demonstrate the simplest possible function: output a single line of text, "Hello World!"

The first step to testing out this default Activity is to build or compile your project. Building instructs Eclipse to collect all of your source files and external resources, evaluate for syntax and reference errors, and then assemble this into code readable by an Android powered device.

ANDROID DOES

By default, Eclipse automatically builds your project with every save. This ensures that your last save also keeps the executable version of your app up-to-date, but it can be processor intensive. If you do not wish your app rebuilt with every comment you save, change the build settings from the Project menu.

By this point, Eclipse has already built your project, and you can actually run the app. Eclipse and most other development environments offer two ways to run compiled code: run and debug. Run, as you might expect, simply executes the code exactly as written and either succeeds or fails depending on what you have written.

Debug, on the other hand, runs the code anticipating errors and traps these errors for you. Debugging enables you to pinpoint the exact point of failure in your app, and it is often your first line of defense when troubleshooting problems with your code.

Whether debugging or not, Android apps need a target device to actually run. This can either be your physical Android powered phone, tablet, or TV, or this can be an emulated Android environment.

Meet the Emulator

To run an app in an emulated Android environment, you must create an Android Virtual Device (AVD). Fortunately, Eclipse makes this a simple process.

1. Right-click **BasicRecipes** and select **Run As > Android Application.** See Figure 2.7.

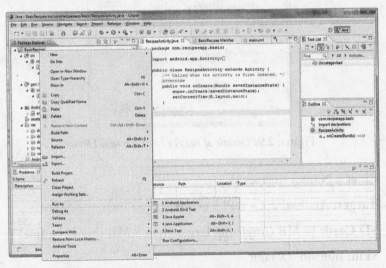

Figure 2.7: *Run your app by opening up the Run As menu.*

2. If you have never created an AVD before, Eclipse will ask if you want to add a device. See Figure 2.8.

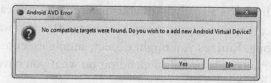

Figure 2.8: *Creating an Android Virtual Device.*

3. The Android SDK and AVD Manager Screen displays. Click **New....** See Figure 2.9.

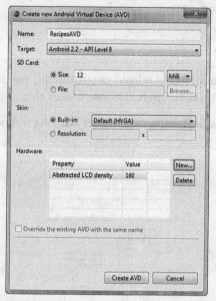

Figure 2.9: *Details of the Android Virtual Device.*

4. Enter a **Name:** RecipesAVD

   Target: Android 2.2 – API Level 8

   SD Card: 12 MB (Anything larger than 8 MB should be fine)

   Skin: Built-in: Default

   Hardware: Leave the defaults.

You can actually add hardware features later by clicking **Hardware > New...** to emulate almost any real world hardware component from Accelerometer to GPS.

   Click **Create AVD** to finish.

5. Click on the **RecipesAVD** and click on **Start....**

6. Click **Launch** and your emulator will display shortly. See Figure 2.10.

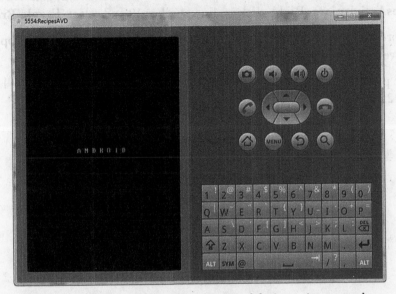

Figure 2.10: *The Android emulator, a virtual device environment where you can test out your apps.*

7. Close the Android SDK and AVD Manager Window.

8. Select the **RecipesAVD** name and click **OK.**

The Eclipse console view should be active by default, and you can watch the progress of your app as it is transferred to the emulator, installed, and launched. Finally, you can see the output of your first Android app.

You have officially created your first Android app from start to finish.

Debugging with the Emulator

Now that you can run your app in the emulator, you can also debug your app in the emulator. Hello World, by design, does not have much room for bugs; after all, it is designed to be as short and simple as possible. You can still debug your app, even if you think it is foolproof by using breakpoints. Breakpoints act as stop signs for the debugger: the debugger will pause execution of the code whenever it sees a breakpoint allowing you to inspect the execution of your code line by line.

From your Package Explorer view, expand your /src folder and double-click **RecipesActivity.java.** Right-click on line number 10 and select **Toggle Breakpoint.**

As before, right-click on your project but select **Debug As > Android Application.** You have already created your AVD, so the Android emulator launches.

Eclipse prompts you to switch into the Debug Perspective. Click **Yes,** as shown in Figure 2.11.

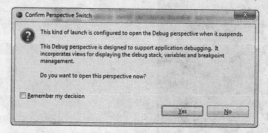

Figure 2.11: *Switching to Debug Perspective.*

The Eclipse debugger has stopped your app from executing at the breakpoint you set. Hit the **F8** key on your keyboard to continue executing the app and display Hello World, or hit **F5** to step into the app and execute it line by line.

Making the Device Connection

While the integrated Eclipse debugger is useful and feature rich, for advanced interaction with the emulated Android environment, developers want to simulate phone calls, SMS text messages, explore the file system, and monitor application threads. Enter the *Dalvik Debug Monitor Service (DDMS)* perspective. The DDMS perspective provides:

- Screen captures
- File/directory access
- Task and thread management
- Logging services
- Direct emulator interaction

Eclipse projects are not required to use the DDMS perspective. Once you have created your AVD, you can start the emulator from the main menu **Window > Android SDK and AVD Manager** and then **Window > Open Perspective > Other > DDMS** (on Windows) or **Window > Open Perspective > DDMS** (on Mac).

> **DEFINITION**
>
> **Dalvik Debug Monitor Service (DDMS)** acts as a middle man between your real or emulated device and Eclipse. Android allows for every process to be monitored in debug mode, and DDMS provides the infrastructure to both passively listen to Android apps as well as actively communicate with them.

The most obvious advantage to using DDMS is the level of simulated hardware interaction it enables. With DDMS, you can spoof location information, sending GPS coordinates to your device. It also enables voice and data signal simulation, interaction with a simulated camera, and screenshots of the device display.

While the DDMS perspective offers advanced features not possible in the standard Eclipse debug perspective, it cannot overcome some of the principle limitations of running Android in an emulated environment. These are some of the practical limitations of an emulated device:

- The emulator is not a real device. In generic ways, it can recreate the behavior of a physical phone or TV, but existing consumer devices are too varied to be represented by a single emulator.

- All signal information is simulated. Proximity sensors, location data, Wi-Fi and tower connection, accelerometer, calls, and SMS messages are entirely simulated.

- Peripheral support is not yet fully implemented. This includes Bluetooth and camera support, as well as some aspects of USB support.

As versatile and robust as the emulator is, it cannot replace testing with a physical Android device.

Congratulations! You have taken the Android development toolkit and used it to create and run an application on an emulated Android environment, running Hello World in standard and debug mode. Taking your project forward into the next chapter, you will leave Hello World behind for more sophisticated and exciting development opportunities in Android OS.

The Least You Need to Know

- Launch Eclipse, then select **File** > **New** > **Project** and select **Android** > **Android Project** to start a new project with a default Activity.

- The Package Explorer view in Eclipse maintains the list of the project files relevant to your app.

- Use an Android Virtual Device (AVD) to run an emulated Android environment.

- Run and debug your apps by launching **Run As** or **Debug As** > **Android Application.**

- The DDMS perspective enables you to simulate real world interaction with your emulated device.

Crafting the Layout

In This Chapter

- App life cycle overview
- Designing the layout of your app
- Adding buttons and labels to your app
- Handling device rotation for your app
- Creating an app icon

Perusing the Android Market, apps vary dramatically in complexity. Google's Shopper app sits at one end of the spectrum: a single text box and a camera icon on a plain white backdrop. Take a picture of a barcode, and you get comparison shopping results in an itemized list. At the other end of the spectrum, the Layer app takes a live video stream from the Android camera and overlays Wikipedia articles, Twitter feeds, and classified ads using GPS to find the relevant data to your location.

You probably want your first apps to be somewhere in the middle of these two ends of the design spectrum. It can be helpful to observe what design decisions other developers have made and whether those are successful. Sometimes simple is key; sometimes not.

By the end of this chapter, you will have created the basic layout of the Recipe app you will be improving throughout the book. In short time, you will have given your app a home screen icon and implemented some text, labels, and buttons into your app.

Application Fundamentals

Before diving head first into design layouts, it is important to pause and consider how design fits into the app as a whole. There are a few *classes* in the Android SDK that are relevant to this question and worth quickly noting here. You will explore these again in greater depth later, but here is a brief introduction:

> **DEFINITION**
>
> A **class** is a Java concept that defines a blueprint for creating an object. The class describes the attributes and methods available about each instance of an object. In this way, a class defines the state and behavior of an object.

- **Activity class.** An *activity* is something a user can do, some single, specific task. Activities usually require user interaction, so the Activity class automatically creates a window to render a user experience. The two most common methods associated with an activity are onCreate() when the user launches the activity, and onPause() when the user switches to another activity.

- **Intent class.** An *intent* is a description of an action or an operation to perform. It can be used to start an activity, start or communicate with a service, or to interact with a broadcast receiver. Almost all actions in Android depend on intents, from "Make a call" or "Search for a song" to "Launch my app".

- **Service class.** A *service* is an action or task that executes in the background, typically not visible by the user. Services allow long-running tasks to continue executing even if the activity is paused, or they can allow continued communication with other apps. Using the Google Maps app, you can get voice-activated turn-by-turn directions across the country. Fortunately, once started, the Maps service ensures directions will continue to speak even after you have switched apps.

- **Content Provider class.** A *content provider* is the best way to communicate information between applications. Google implemented contact providers for most of the default apps: contacts, images, audio, and video. To read or write data into a user's photo library, you must simply use the appropriate Content Provider and request permission, if necessary.

These four classes form the cornerstones of your app development in Android. Most of your work in designing an app focuses on specific activities you create to display content to the user. Activities have their own *life cycle* which is represented in Figure 3.1.

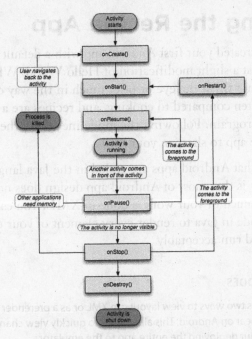

Figure 3.1: *The life cycle of your app.*

"Portions of this page are reproduced from work created and shared by the Android Open Source Project and used according to terms described in the Creative Commons 2.5 Attribution License." You can find it at http://developer.android.com/reference/android/app/Activity.html.

The four ovals represent the primary states your activity can occupy: Started, Running, Killed, or Shutdown. The rectangles represent possible methods to call during these states. You will spend most of your development time thinking about your app in its running state; that is when someone is actively using your app. It is important to remember that users switch back and forth between apps, and that your app moves through its entire application life cycle many times.

DEFINITION

An object's **life cycle** is the time between when the object is created or started and when it is ended or destroyed.

Reintroducing the Recipe App

In Chapter 2, you created your first Android app with a default activity, RecipeActivity, which was really just a slight modification of Hello World. While well and good, Hello World does not thrill the eye or offer much in the way of user experience. Programming is often compared to cooking, and recipes are a great way to think about learning to program. Following the guidelines that others have set forth, you will create a Recipe app to sharpen your skills.

You already know that Android apps are coded in the Java language, but what you might not yet know is that most of Android app design does not need to be written in Java at all. Rather, much of your work is done in XML. You can, of course, choose to write procedural code in Java to render every element of your design; and this would be build, install, and run acceptably.

ANDROID DOES

Eclipse offers two ways to view layouts: in XML or as a prerender of the layout as it will appear on Android. This allows you to quickly view changes to your layout without deploying the entire app to the emulator.

Google has optioned another, easier path: write as much as possible in simple XML. XML is less likely to change between versions of the OS, it is easy to read and modify, and when built, it is compressed and optimized for the target Android device.

Laying Out the Welcome Screen

In Eclipse, open (or create) the BasicRecipes project from Chapter 2. Also take the time to open your Android Emulator, RecipesAVD. This allows you to quickly see the changes you make to your app as you write your code. The emulator consumes few resources and can stay open as long as you are in Eclipse. If you run your app now, you should see the familiar, "Hello World, RecipesActivity!"

Activities are responsible for displaying the user experience. If you explore the RecipesActivity.java, you should see:

```
BasicRecipes/src/com.recipeseapp.basic/RecipeActivity.java
package com.recipesapp.basic;
import android.app.Activity;
import android.os.Bundle;
```

```
public class RecipesActivity extends Activity {
    /** Called when the activity is first created. */
    @Override
    public void onCreate(Bundle savedInstanceState) {
        super.onCreate(savedInstanceState);
        setContentView(R.layout.main);
    }
}
```

From the activity life cycle, you know that onCreate() happens when the activity is called. The only remaining method called is setContentView(). This method fills an activity's display with the contents of an Android view. In this case, R.layout.main references main.xml in your project /res/layout folder, as shown in Figure 3.2.

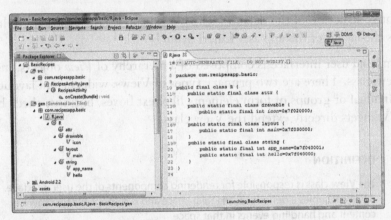

Figure 3.2: *The R.layout of your app.*

The R class provides quick references to the resources available in your project. Every Android project receives an **R.java** file in the /gen folder, and your project's /gen folder is automatically updated with every change you make to your resources.

Double-click to open main.xml from /res/layout/main.xml, then click on the **main. xml** tab below the graphics. It may take a moment to load:

```
<?xml version="1.0" encoding="utf-8"?>
<LinearLayout xmlns:android="http://schemas.android.com/apk/res/
android"
    android:orientation="vertical"
    android:layout_width="fill_parent"
    android:layout_height="fill_parent"
    >
```

continues

```
<TextView
    android:layout_width="fill_parent"
    android:layout_height="wrap_content"
    android:text="@string/hello"
    />
</LinearLayout>
```

Now open **strings.xml** from **/res/values/strings.xml**. Again, tap the **strings.xml** tab once you open it:

```
<?xml version="1.0" encoding="utf-8"?>
<resources>
    <string name="hello">Hello World, RecipesActivity!</string>
    <string name="app_name">The Basic Recipes</string>
</resources>
```

The entire structure of the default Hello World app is now exposed. Main.xml defines the layout of the app as a LinerLayout.

The Android user interface is composed of a hierarchy of *View* objects, instances of the View class. There are two primary types of Views: widgets and layouts. Widgets are individual or groups of form elements like text boxes, buttons, check boxes, and labels. Widgets directly extend the View class.

DEFINITION

The **View** class is responsible for rendering components of the user interface. A View occupies a defined rectangular space and is responsible for drawing content and handling events in that space.

Layouts provide a way to organize form elements or widgets for display. Layouts indirectly extend the View class through the ViewGroup class. Layouts primarily define the relative or absolute position of their children, and layouts can be nested inside each other.

The most common layouts in Android are:

- **FrameLayout:** Defines a single area of the screen to block out for display. Children are attached to the top-left, and the most recently attached is on top. This can be useful for tabbed interfaces.

- **LinearLayout:** Renders its children as a single column (vertically) or a single row (horizontally).

- **RelativeLayout:** Positions children relative (below, above, left, or right) to the parent or other children.

- **TableLayout:** Displays children in rows and columns. The TableRow class defines each row, and each cell is its own view.

- **AbsoluteLayout:** A layout of child elements in absolute x/y pixel coordinates. This layout is inflexible and least commonly used.

GOOGLE IT

Every Android class holds more properties and methods than can be covered in a single text. When in doubt about using unfamiliar XML tags, consult http://android.developers.com or Google it.

By analyzing the two XML files noted previously, it should be clear that you have defined a LinearLayout, oriented vertically, which fills all available screen space and displays the text of string, tag name="hello" to display "Hello World, RecipesActivity!" on screen.

Now it is time to replace this skeleton with some more meaningful text and buttons. Here is the XML syntax for buttons:

```
<Button
        android:id="@+id/my_button"
android:layout_width="wrap_content"
        android:layout_height="wrap_content"
android:text="@string/button_label" />
```

Let's break down the code here:

- **id** provides a unique identifier for the button to provide access to the object later. This is optional but recommended.

- **layout_width** defines the width of the object in relative or absolute terms.

- **layout_height** defines the object height in relative or absolute terms.

- **text** defines the content to display in the button.

You can now create a complete Welcome menu with ease.

In main.xml, enter:

```xml
<?xml version="1.0" encoding="utf-8"?>
<LinearLayout xmlns:android="http://schemas.android.com/apk/res/
    android"
    android:orientation="vertical"
    android:layout_width="fill_parent"
    android:layout_height="fill_parent"
    >
<TextView
    android:id="@+id/welcome_title"
    android:layout_width="fill_parent"
    android:layout_height="wrap_content"
    android:gravity="center"
    android:textStyle="bold"
    android:text="@string/welcome_title"
    />
<Button
    android:id="@+id/search_button"
    android:layout_width="fill_parent"
    android:layout_height="wrap_content"
    android:text="@string/search_button"
    />
<Button
    android:id="@+id/new_button"
    android:layout_width="fill_parent"
    android:layout_height="wrap_content"
    android:text="@string/new_button"
    />
<Button
    android:id="@+id/help_button"
    android:layout_width="fill_parent"
    android:layout_height="wrap_content"
    android:text="@string/help_button"
    />
<Button
    android:id="@+id/exit_button"
    android:layout_width="fill_parent"
    android:layout_height="wrap_content"
    android:text="@string/exit_button"
    />
</LinearLayout>
```

In strings.xml, insert:

```xml
<?xml version="1.0" encoding="utf-8"?>
<resources>
        <string name="app_name">Recipe Basics</string>
        <string name="welcome_title">Welcome to Your Recipes!</string>
        <string name="search_button">Find a Recipe</string>
        <string name="new_button">Create New Recipe</string>
        <string name="help_button">Help</string>
        <string name="exit_button">Quit</string>
</resources>
```

By default, Eclipse builds your project automatically on save. You can immediately run your project to see the new Recipe Welcome screen: Right-click on the project name **BasicRecipes,** do **Run As > Android Application.** The emulator will run the latest version of your app, as Figure 3.3 illustrates.

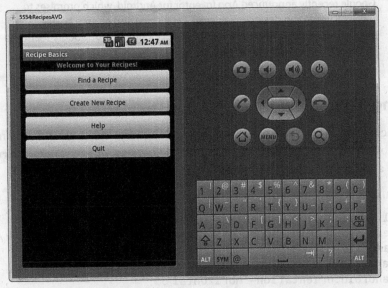

Figure 3.3: *The emulator running your recipe app.*

The emulator renders in portrait mode by default, but you can switch into landscape mode by hitting **Ctrl + F11.** You should make a point of testing your app in all viewing modes with each major change to ensure it always looks correct.

ANDROID DOES

You should notice that as you begin typing each of these XML elements, Eclipse offers suggestions to autocomplete your text with possible values. Use these! This is one of the biggest advantages of your IDE: code assistance. Stuck in XML? Type **android:** and Eclipse shows you every possible tag.

If you have previously developed for desktop or mobile platforms that include graphical form designers, this first experience in Android user interface design may seem like a step sideways rather than forward. However, if you have spent any time developing for the web, this approach to design will quickly feel intuitive and routine. Writing XML for Android is quite similar to writing HTML for the web, and with a little patience, it soon seems second nature.

GOOGLE IT

Google recently announced App Inventor for Android, which promises to deliver a visual app design experience. DroidDraw also offers a basic, no frills visual design utility. Both tools are still in development, but Google them for more information about availability.

Expanding Your View

The Recipe app Welcome screen has come a long way, but much more can be done to improve its aesthetic. Without changing the content of the screen, rearrange the buttons and text to center the focus:

In main.xml, replace the current code with the following:

```
<?xml version="1.0" encoding="utf-8"?>
<LinearLayout xmlns:android="http://schemas.android.com/apk/res/
    ➥android"
    android:id="@+id/main_layout"
    android:orientation="horizontal"
    android:layout_width="fill_parent"
    android:layout_height="fill_parent"
    android:padding="10dp"
    >
    <LinearLayout
    android:id="@+id/mainsub_layout"
    android:orientation="vertical"
    android:layout_width="wrap_content"
```

```
        android:layout_height="wrap_content"
        android:layout_gravity="center"
        android:paddingTop="20dp"
        android:paddingBottom="20dp"
        >
<TextView
        android:id="@+id/welcome_title"
android:layout_width="fill_parent"
        android:layout_height="wrap_content"
        android:gravity="center"
        android:textSize="20dp"
        android:textStyle="bold"
        android:text="@string/welcome_title"
        />
<RelativeLayout
        android:id="@+id/button_layout"
        android:layout_width="wrap_content"
        android:layout_height="wrap_content"
        >
<Button
        android:id="@+id/search_button"
        android:layout_width="fill_parent"
        android:layout_height="wrap_content"
        android:text="@string/search_button"
        />
<Button
        android:id="@+id/new_button"
        android:layout_width="fill_parent"
        android:layout_height="wrap_content"
        android:layout_below="@+id/search_button"
        android:layout_alignLeft="@+id/search_button"
        android:text="@string/new_button"
        />
<Button
        android:id="@+id/help_button"
        android:layout_width="fill_parent"
        android:layout_height="wrap_content"
        android:layout_below="@+id/new_button"
        android:layout_alignLeft="@+id/new_button"
        android:text="@string/help_button"
        />
<Button
        android:id="@+id/exit_button"
        android:layout_width="fill_parent"
        android:layout_height="wrap_content"
```

continues

```
            android:layout_below="@+id/help_button"
            android:layout_alignLeft="@+id/help_button"
            android:text="@string/exit_button"
            />
    </RelativeLayout>
    </LinearLayout>
</LinearLayout>
```

Now run the app by right-clicking the **BasicRecipes** project name and doing **Run As > Android Application.**

The buttons in the Welcome screen are now neatly centered vertically in the page, and the title is larger and more readable. Much more has changed in the underlying. This uses layouts nested three levels deep: LinearLayout > LinearLayout > RelativeLayout. The outermost layout defines a margin or pad for the entire window with the dp unit. See Figure 3.4.

Figure 3.4: *Your recipe app with the centered menu screen.*

For the purpose of defining size, Android supports these units of measure:

By Height and Width and Breadth

Unit	Name	Definition
dp	Density Independent Pixels	A unit relative to the pixel density of the screen. Sometimes referenced as dip.
sp	Scale Independent Pixels	A relative unit like dp but scaled against individual users' font size preferences.
pt	Points	$1/72$ of an inch.
mm	Millimeter	Actual size as measured by a metric ruler.
in	Inches	Actual size as measured by an Imperial ruler.
px	Pixels	Number of pixels on display.

In normal development, it is preferable to use the relative sizes dp and wp to specify height and width as Android will scale these sizes in proportion to the actual device display. In this way, you can design your activities to render consistently from screens very small to quite large.

In portrait mode, with the device held vertically, the Welcome screen looks good, but what happens when the user rotates the screen? Remember, you can tap **Ctrl+F11** to flip the orientation from vertical to horizontal. See Figure 3.5.

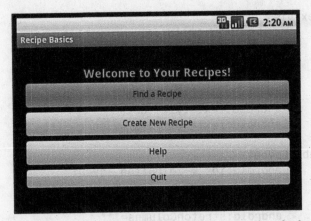

Figure 3.5: *The app screen also centers the buttons when the device is held horizontally in landscape mode.*

The Quit button is now half as small as the other buttons. There are two ways to solve this issue: tune the XML to render in both orientations, or create two layouts. If you were to add more text or buttons to the Welcome screen, you can easily imagine how formatting a single window for multiple orientations could become nightmarish. Fortunately, Android anticipates this dilemma and offers a simple solution: landscape layout definitions.

First, go to **File > New > File,** select the folder **Res** and add a new folder, **layout-land.** Now, create a main.xml in BasicRecipes/res/layout-land as:

```xml
<?xml version="1.0" encoding="utf-8"?>
<LinearLayout xmlns:android="http://schemas.android.com/apk/res/
➥android"
        android:id="@+id/main_layout"
         android:orientation="horizontal"
        android:layout_width="fill_parent"
        android:layout_height="fill_parent"
        android:padding="10dp"
        >
    <LinearLayout
    android:id="@+id/mainsub_layout"
    android:orientation="vertical"
    android:layout_width="wrap_content"
    android:layout_height="wrap_content"
    android:layout_gravity="center"
    android:paddingTop="10dp"
    android:paddingBottom="10dp"
    >
    <TextView
            android:id="@+id/welcome_title"
            android:layout_gravity="center"
            android:layout_width="wrap_content"
            android:layout_height="wrap_content"
            android:textSize="25dp"
            android:textStyle="bold"
            android:text="@string/welcome_title"
    />
    <TableLayout
            android:id="@+id/button_layout"
            android:layout_width="wrap_content"
            android:layout_height="wrap_content"
            android:stretchColumns="*"
    >
            <TableRow>
```

```
<Button
        android:id="@+id/search_button"
        android:text="@string/search_button"
/>
<Button
        android:id="@+id/new_button"
        android:text="@string/new_button"
/>
</TableRow>
<TableRow>
<Button
        android:id="@+id/help_button"
        android:text="@string/help_button"
/>
<Button
        android:id="@+id/exit_button"
        android:text="@string/exit_button"
/>
</TableRow>
</TableLayout>
</LinearLayout>
</LinearLayout>
```

Using a TableLayout, landscape mode (see Figure 3.6) now renders the buttons in ordered column pairs. Alternate layout files not only accommodate for all possible orientations of the Android device, but also for localization by language and region.

Figure 3.6: *Your recipe app using the column-based TableLayout mode.*

Formatting for All Displays

You can specify resource folders for many more display conditions than portrait and landscape. By creating new resource folders and layout files, you can customize layouts by language, time of day, and screen size. Some of the common qualifiers include the following.

Resource Folder Naming Conventions

Condition	Values	Description
Language and region	en, fr, es	Localize your content by region.
Screen size	small, normal, large	Screen size by pixel density (QVGA, HVGA, WVGA, and so on) and aspect ratio.
Screen aspect	long, notlong	The aspect ratio of the screen, where long is wider. This is separate from screen orientation.
Screen orientation	port, land	Portrait or landscape.
Dock mode	car, desk	If docked, whether to a computer or a car.
Night mode	night, notnight	Nighttime or not.
Screen pixel density (dpi)	ldpi, mdpi, hdpi, nodpi	Low, medium, high, or any pixel density.
Touchscreen type	notouch, stylus, finger	The touch capability of the display.
Keyboard availability	keysexposed, keyssoft	Whether the device has a hardware or a software keyboard.
Navigation key availability	navexposed, navhidden	Whether navigational keys are accessible.
System Version (API Level)	v7, v8	The supported Android OS (v7 matched Android 2.1, v8 Android 2.2).

These conditional qualifiers allow you to be quite specific in defining how you want your app to look and feel based on a wide variety of conditions. Qualifiers must be used in the order they are listed in this table. For example, /res/en-land-night would specify resources for English speaking users in landscape mode at night, while /res/land-en would be invalid—language must come before orientation.

Your Icon

Icons identify your app in menus and on the home screen. While it may be difficult to perfectly craft the right icon for your app, it is easy to implement once you have a design. Android classifies icons in three categories:

Type	Low density (ldpi)	Medium density (mdpi)	High density (hdpi)
Launcher/Menu	36×36 px	48×48 px	72×72 px
Status Bar	24×24 px	32×32 px	48×48 px

Expanding /res in Package Explorer will reveal three default folders: drawable-hdpi, drawable-mdpi, and drawable-ldpi. Each contains a single file, icon.png. To create your own, unique app icon, you need only replace this icon image file with an icon of your own for each pixel resolution.

GOOGLE IT

The Android developer site includes extensive guidelines for designing icons, including a comprehensive icon template package which is free to use. Numerous other developer resources online provide free images and templates for your reuse.

Icons for the recipe project in this text were taken from Androidicons.com, which you are free to use yourself in implementing your own app icons. After downloading or creating three versions of your app icon for low, medium, and high density display, you can drag them into the respective resource folders. Follow these steps:

1. Open your file explorer.

2. Click and drag the 36 × 36 px icon into /res/drawable-ldpi.

3. Elect to copy the file into the project, as shown in Figure 3.7.

4. Click and drag the 48 × 48 px icon into /res/drawable-mdpi.

5. Click and drag the 72 × 72 px icon into /res/drawable-hdpi.

6. Check to make sure your folder tree looks like Figure 3.8.

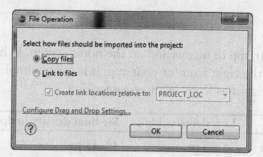

Figure 3.7: *Copying the icon files into your project.*

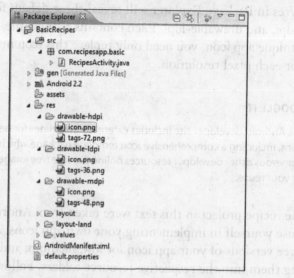

Figure 3.8: *The folder tree now includes your new app icons.*

7. Either rename the existing icon.png (icon_old.png for example) files or delete them from your project.

8. Select your new icon files and hit **F2** to rename them as icon.png.

9. Run the project in the emulator to see the new application icon. See Figure 3.9.

Figure 3.9: *The Android emulator home screen with your new icon!*

Implementing your own icon is as simple as finding or creating the correct image files and importing them into your project.

Introducing Your First Error

Your path as a developer is not limited to the example code and XML provided in this text, and you will eventually meet Android's friendly but unhelpful fatal error page.

While none of the code examples provided so far will produce such an error, you can test the process yourself by removing a required XML tag from inside one of your layout definitions, for example the layout_height tag. Save the change and run the app against your emulator to see the error. See Figure 3.10.

The Android SDK provides a simple and direct path to diagnosing many of these errors by providing extensive logging. To view the output of the emulator log, open the LogCat view (see Figure 3.11) in Eclipse from **Window > Show View > Other > Android > LogCat.**

Figure 3.10: *An Android emulator error page.*

Implementing your own icon is as simple as finding the correct image files and importing them.

Introducing Your First E

Very rarely is a developer not limited to just sample code and XML provided in chapters, and you will eventually need to deal with ugly but helpful fatal error pages.

While none of the code examples provided here will produce such an error, you can test the process yourself by inducing a required setting from inside one of your layout definitions, for example the layout sheet. Save the change and run the app against your emulator to see the error. See Figure 3.10.

The Android SDK provides a tool used to help solve diagnosing many of these errors by providing critical information from the emulator log, often the LogCat view (see Figure 3.11) from the menu Window > Show View > Other > Android...

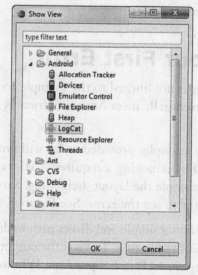

Figure 3.11: *The LogCat view allows you to diagnose errors easier.*

If you were to scroll through the log looking for the culprit behind this error (see Figure 3.12), you would eventually find: "Caused by: java.lang.RuntimeException: Binary XML file line #18: You must supply a layout_height attribute."

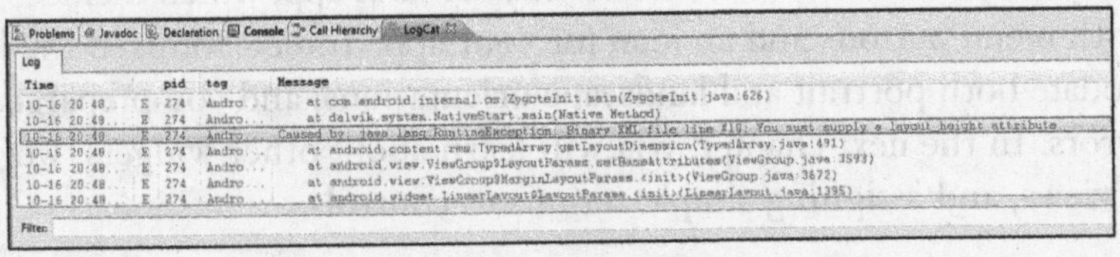

Figure 3.12: *The LogCat view showing you errors in the code.*

There is another path to debugging your apps, which is to run the app through the debugger. In order to do this, you must first enable debugging in your app by setting debuggable="true" in the AndroidManifest.xml file. This can be accomplished by manually editing the source XML setting (see Figure 3.13):

```
<application
android:icon="@drawable/icon"
android:label="@string/app_name"
android:debuggable="true"
>
```

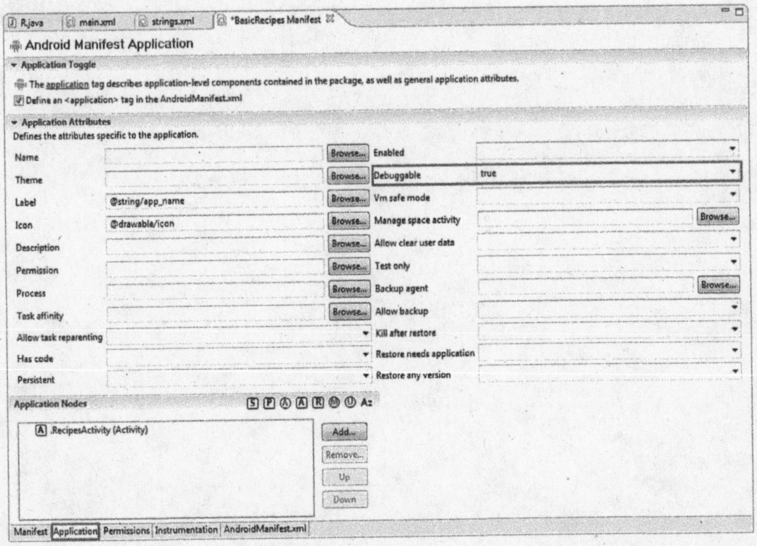

Figure 3.13: *Eclipse also offers a visual assistant to accomplish the same task.*

Now you can run your app through the debugger by selecting **Run > Debug As > Android Application.** The debug perspective will open, and you can begin trouble-shooting your app.

In this chapter, you created the skeleton of your first app, which includes a Welcome screen with menu buttons and an icon for your app. You customized the layout to accommodate both portrait and landscape orientations, and you prepared yourself to debug errors. In the next chapter, you will learn about other widgets, working with multiple views, and assigning simple actions to controls.

The Least You Need to Know

- Android user interfaces belong to Activities.
- Each activity represents a set of views to render a display to the user.
- Views can be nested within each other to customize the layout of an activity.
- Resources stored in your /res folder define alternate layouts for different regions, environments, and displays.
- Use CatLog and run your app in debug mode to troubleshoot errors.

Finding More Activities

4

In This Chapter

- Link buttons with events
- Use styles and themes that increase your app appeal
- Create option menus
- Collect user preferences

In your app journeys through the Android Market, you will undoubtedly find apps that can perform just about any task. From flashlights to scientific calculators, you will find apps that do very little and others that do everything short of making dinner. In fact, there are only two design criteria for any application: there must be something to see and something to do.

In Chapter 3, you met the first requirement by creating some visual content with buttons and text. If this were a flashlight app, the user interface and the functionality of the app could be one and the same: generate a bright, white screen. The recipe app is a little bit more ambitious, however, so more work needs to be done to make this app truly functional. In this chapter, you continue to enrich the interface and add some actions to the app.

Improving Your Recipe App

In Chapters 2 and 3, you created the BasicRecipe project and implemented a single activity, the RecipesActivity. Beyond showing a menu of buttons, the app does not perform any additional function (see Figure 4.1). It is time to take a step back and ask two fundamental questions: "Who will use this app?" and "What will this app do?"

Figure 4.1: *Your current recipe app.*

Answering the first question often informs the second. Users are likely adults who either like to cook or are learning to cook. Given that, the app should probably allow users to create and modify recipes, search and share them with friends, and take pictures of finished dishes and ingredients. These should be documented as the requirements of your app. Reference and revise them as you complete code.

Introduction to Refactoring

Open or create the BasicRecipes project in Eclipse (as shown in Chapter 2) and look at RecipesActivity in Package Explorer. The name of this activity does not quite capture how it is actually used. This activity is a menu on the first screen to open. As you add new activities to your project, you can save yourself a great deal of head scratching by naming your objects well.

Changing a class name is not as easy as just renaming the class itself, because other objects may already be referencing it by name. Fortunately, Eclipse solves this problem with the concept of *refactoring*. In this context, refactoring changes the name of the activity and updates all references to the activity with the new name.

DEFINITION

Refactoring is the process of improving source without changing its expected behavior. Reasons to refactor can include improving readability, decreasing complexity, or improving maintainability of code.

Here's how we'll refactor our recipe program:

1. In Package Explorer, click **RecipesActivity.java.**

2. Select **Refactor > Rename.**

3. Enter **MainMenu** as the new name (see Figure 4.2) and click **Finish.**

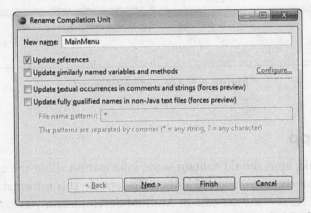

Figure 4.2: *Renaming a section of your program, which is called "refactoring."*

4. Eclipse has now updated all uses of the name, except in the Android Manifest. See Figure 4.3.

If you run the app now, you will find that it behaves exactly as it did before. Only you, the designer, would ever know that you successfully refactored your app.

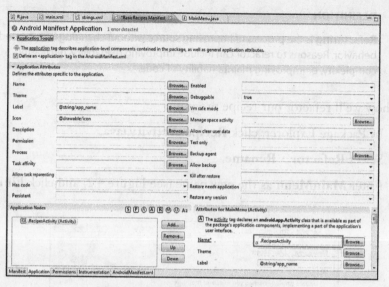

Figure 4.3: *The Android Manifest.*

About Your App

As a general rule, apps should contain some information about the app itself, including version number, name, and the author. To provide this information, you create an About page. As is often the case in Android development, you can accomplish this multiple ways. You can either:

- Create a new activity and run it.
- Use an existing class, such as AlertDialog or Dialog, and show it.

Both options are worth exploring. In the first case, several steps are involved:

1. Create a new activity class, About.

2. Create a new XML doc for the About activity layout.

3. Add relevant content to the strings.xml resource.

4. Add an About button to the MainMenu activity layout.

5. Modify the MainMenu class to call this activity when the About button is clicked.

When you created the BasicRecipes project, Eclipse offered an option to create a default activity. Unfortunately, the process to create new activities is still a bit manual.

Updating the Android Manifest

Whenever you add new classes to your project, the Android Manifest needs to be updated. While you can implement any number of features in your project inside Eclipse, Android has no way to know about your new objects unless they are explicitly defined in the Manifest.

PITFALL

Eclipse does not generate build errors if content is missing from your manifest. If your manifest is incorrect, errors will not surface until you try to run the app.

By starting with the Android Manifest in the process of adding a new activity, you will see an error in the manifest file until the activity is implemented.

There are two ways to implement. You can type the change directly into the code:

1. Open **AndroidManifest.xml.**
2. If you are already comfortable working with XML, insert a new activity:
   <activity android:name=".About"></activity>.

Alternatively, you can use the visual editor:

1. Open **AndroidManifest.xml.**
2. Using the visual editor works as well and guarantees the XML will be formatted correctly. From the manifest tabs, select **Application > Application Nodes > Add.**
3. Select **Create new element** at the top level. See Figure 4.4.
4. Click **Activity,** and then click **OK.**
5. Give the activity a name of .About under Attributes for Activity.

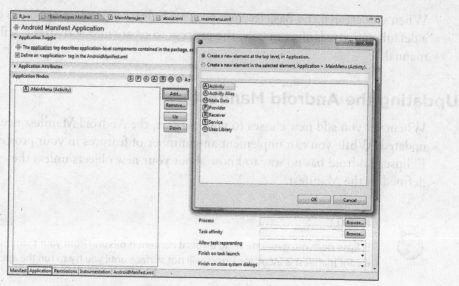

Figure 4.4: *Adding a new activity to your program.*

Adding an Activity

Now that the manifest knows to look for the new activity, create the corresponding class in Eclipse:

1. Right-click on the **com.recipesapp.basic** node in the Package Explorer.

2. Select **File > New > Class** (on Windows) or **New > Class** (on Mac).

3. Enter these values (see Figure 4.5):

   Name: **About**

   Superclass: **android.app.Activity**

   Check **Constructors from superclass**

4. Click **Finish** when you're done and it will create the About.java class.

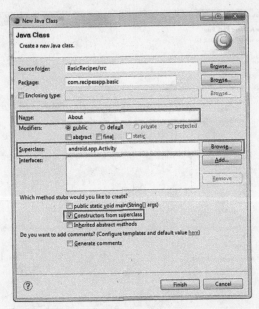

Figure 4.5: *Adding details to your new activity.*

5. Edit the new About.java class so that it contains:

```
package com.recipesapp.basic;
import android.app.Activity;
import android.os.Bundle;
public class About extends Activity {
    @Override
    public void onCreate(Bundle savedInstanceState) {
        super.onCreate(savedInstanceState);
        setContentView(R.layout.about);
    }
}
```

Creating the Layout

Your activity is ready to render, but it needs content to display. About.xml does not exist yet, so you need to create it:

1. Select **File > New > Android XML File.** See Figure 4.6.

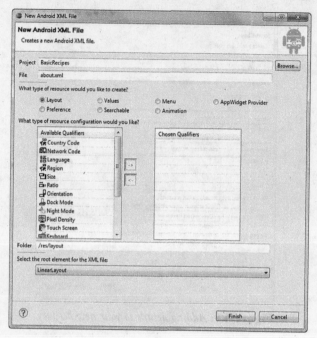

Figure 4.6: *Creating a new Android XML file.*

2. Go to the /res/layout directory and define the XML file as a new Layout resource:

   File: **about.xml**

   Resource type: **Layout**

   Folder: **/res/layout**

   XML root: **LinearLayout**

3. Open **About.xml** and define a simple layout to display some text:

```xml
<?xml version="1.0" encoding="utf-8"?>
<LinearLayout xmlns:android="http://schemas.android.com/apk/res/
  ➥android"
      android:id="@+id/about_layout"
      android:layout_width="wrap_content"
      android:layout_height="wrap_content"
      android:padding="10dp"
    >
```

```
<TextView
        android:id="@+id/about_summary"
        android:layout_width="fill_parent"
        android:layout_height="wrap_content"
        android:gravity="center"
        android:textSize="20dp"
        android:text="@string/about_summary"
    />
</LinearLayout>
```

4. Define the text values in strings.xml, and then review Figure 4.7:

```
<string name="about_header">About Simply Recipes</string>
<string name="about_summary">Simply Recipes allows you to
manage your recipes and share them with friends.</string>
<string name="main_about_button">About</string>
```

Figure 4.7: *The beginning of your About page.*

The basic content of the About page is now in place, but you still need to create an About button and implement the event.

1. Add a button to the layout of main.xml, right after the other buttons and right before the RelativeLayout loop ends:

```
<Button
        android:id="@+id/main_about_button"
        android:layout_width="fill_parent"
        android:layout_height="wrap_content"
        android:layout_below="@+id/exit_button"
        android:layout_alignLeft="@+id/exit_button"
        android:text="@string/main_about_button"
/>
```

2. Add the following text to the layout-land/main.xml file to make sure the About button shows in the 2-column table, right above the </TableLayout> tag:

```
<TableRow>
        <Button
        android:layout_span="2"
        android:id="@+id/main_about_button"
        android:text="@string/main_about_button"
        />
</TableRow>
```

Putting It Together

Finally, it's time to connect the new About activity with the About button. The design goal is to associate a click event on the About button to launch the About activity. While most of the work done to date has involved working with XML, some procedural code has to be written in Java. Start by opening the MainMenu.java class.

First, the necessary methods to handle click events are not accessible in the context of this activity, so you must add them as imports in your class:

```
import android.content.Intent;
import android.view.View;
import android.view.View.OnClickListener;
```

Next, the MainMenu class needs to do something. If you are familiar with Java, you may want to start by defining a new, anonymous class; but all new inner classes consume additional resources. *Interfaces*, on the other hand, are free. To implement the correct interface to use the appropriate methods for this action, modify the class to implement OnClickListener:

```
public class MainMenu extends Activity implements OnClickListener
```

DEFINITION

An **interface** is frequently used to define an abstract type that contains only a set of methods. If a class contains all of these methods, it is said that the class implements the interface.

Now you can write some code to associate an event with the About button. Inside onCreate(), add the following code:

```
View aboutButton = findViewById(R.id.main_about_button);
aboutButton.setOnClickListener(this);
```

When you use the "@+id" syntax to define IDs in XML for your layouts and widgets, the Eclipse ADT automatically includes these into your R class, which makes accessing the content of your views very simple.

The OnClickListener interface defines a single method, onClick(), which means that you need to implement that method and define some action:

```
public void onClick(View thisView) {
        switch (thisView.getId()) {
        case R.id.main_about_button:
        Intent showAbout = new Intent(this, About.class);
        startActivity(showAbout);
        break;
        }
}
```

Your code is complete and ready to compile and run. Your entire MainMenu.java file should now look like:

```
package com.recipesapp.basic;
import android.app.Activity;
import android.os.Bundle;
import android.content.Intent;
import android.view.View;
import android.view.View.OnClickListener;
public class MainMenu extends Activity implements OnClickListener {
    /** Called when the activity is first created. */
    @Override
```

continues

```
public void onCreate(Bundle savedInstanceState) {
    super.onCreate(savedInstanceState);
    setContentView(R.layout.main);

    View aboutButton = findViewById(R.id.main_about_button);
    aboutButton.setOnClickListener(this);
}

public void onClick(View thisView) {
    switch (thisView.getId()) {
    case R.id.main_about_button:
        Intent showAbout = new Intent(this, About.class);
        startActivity(showAbout);
        break;
    }
}

}
```

Run your app in the emulator (see Figure 4.8).

Figure 4.8: *The final About page for the app.*

Your Recipes app now has a functional About page. As you come back to implement
behavior for the other buttons on the Welcome screen, you will repeat many of these
steps.

Styling Your App

The first draft of your About page looks good, but it could use some flair. Much of the process of editing the look and feel of an Android app is quite similar to changing styles in HTML and Cascading Style Sheets (CSS). You can modify the style or theme of the entire app, then modify the aesthetics of each activity, each level of each view, down to the individual widgets.

Selecting a Theme

Android references a collection of visual settings as a *style*. Styles can be applied to Views and ViewGroups. All child objects within the view will inherit the style, but child Views do not. A style's scope is limited to the view in which it is applied. To cascade styles down a view hierarchy, use a *theme*. Themes must be defined in the Android Manifest for their scope.

> **DEFINITION**
>
> A **style** is a template of property definitions including width, height, font color and size, background color, and more.
>
> A **theme** is a style when applied to an entire activity or the whole app.

The Android SDK includes a number of default styles and themes, which you can use immediately. Open **AndroidManifest.xml** and add style tags to the entire app.

Add android:theme="@android:style/Theme.Light" into the <application> definition:

```
<application android:icon="@drawable/icon"
        android:label="@string/app_name"
        android:debuggable="true"
        android:theme="@android:style/Theme.Light">
```

Now change the theme of the About app by adding android:theme="@android:style/Theme.Dialog" into the <activity> definition:

```
<activity
android:name=".About"
android:label="@string/about_header"
android:theme="@android:style/Theme.Dialog"
>
</activity>
```

The Welcome screen now inherits the style of the Light theme (see Figure 4.9), and the About screen now uses the Dialog style (see Figure 4.10).

Figure 4.9: *Your Welcome screen using the Light theme.*

Figure 4.10: *Your About screen with the Dialog style.*

Experiment with the default styles and themes until you find one you like or create your own.

GOOGLE IT

Styles are not well documented by Google. To get a complete list of available styles and themes, the search engine is your friend.

Rolling Your Own Style

Creating your own style is as simple as, you guessed it, creating a style XML document. Create a styles.xml file in your /res/values folder with this content:

```xml
<?xml version="1.0" encoding="utf-8"?>
<resources>
    <style name="MainStyle" parent="@android:style/TextAppearance.
    Medium">
                <item name="android:layout_width">fill_parent</item>
                <item name="android:layout_height">wrap_content</item>
                <item name="android:gravity">center</item>
                <item name="android:textSize">20dp</item>
                <item name="android:textStyle">bold</item>
        </style>
</resources>
```

To implement this style, you need only insert the style into a view:

```xml
<TextView
        android:id="@+id/welcome_title"
        style="@style/MainStyle"
        android:text="@string/welcome_title"
/>
```

Rather than painstakingly creating individual properties for each of your many views, you can craft a few styles to quickly add collections of settings to a view. Available properties vary by class. To see a complete list of style properties by class, visit the Android Developer site. For example, http://developer.android.com/reference/android/widget/TextView.html#lattrs lists all XML attributes available in the TextView class.

Polishing Your App

You now have many of the necessary tools to start building content and functionality in your app. Granted, Simply Recipes does not yet do very much, but you will start remedying that in Chapter 5. Before beginning additional design work, you should familiarize yourself with two core Android features: menus and settings.

Adding a Menu

Menus provide quick access to the most common functions of an app. In a desktop application, menus can be long and expansive. In the mobile space, screen real estate is expensive and menus should be simpler, shorter, and faster. Android offers two types of menus:

- **Context menus.** Accessed from a long-click, when you tap and hold an object on screen. In the Phone app, tapping on a recently dialed number will give you options to call the number or save it to a contact.

- **Options menus.** Accessed by clicking the menu button on the Android device. Notice how this menu changes between the different tabs of the Phone app.

The Simply Recipes app does not yet have any content to apply a context menu, so you will create an options menu. Without reading ahead, try to visualize how to implement this in the recipe app. The menu content will need to exist somewhere as XML and an activity will need to call a method to load the menu.

Create a new Android XML document in Eclipse as resource type menu, in the res/menu/ folder, with this content:

```xml
<?xml version="1.0" encoding="utf-8"?>
<menu
        xmlns:android="http://schemas.android.com/apk/res/android">
        <item
                android:id="@+id/main_menu_new"
                android:title="@string/main_menu_new"
                android:alphabeticShortcut="@string/main_menu_
➥newshortcut"
                android:orderInCategory="1"
                >
        </item>
```

```
        <item
                android:id="@+id/main_menu_search"
                android:title="@string/main_menu_search"
                android:alphabeticShortcut="@string/main_menu_
    ➥searchshortcut"
                android:orderInCategory="2"
                >
        </item>
        <item
                android:id="@+id/main_menu_options"
                android:title="@string/main_menu_options"
                android:alphabeticShortcut="@string/main_menu_
    ➥optionsshortcut"
                android:orderInCategory="3"
                >
        </item>
    </menu>
```

Add the strings into your string definition in res/values/strings.xml:

```
        <string name="main_menu_search">Search</string>
        <string name="main_menu_searchshortcut">Find a Recipe</string>
        <string name="main_menu_new">New</string>
        <string name="main_menu_newshortcut">Create a New Recipe</
    ➥string>
        <string name="main_menu_options">Options</string>
        <string name="main_menu_optionsshortcut">Your Preferences</
    ➥string>
```

Just as with the About activity, you now need to hook the menu into the MainMenu activity. MainMenu cannot reference the right menu methods based on its current imports, so you will need to add a few more. Open MainMenu.java and add:

```
        @Override
        public boolean onCreateOptionsMenu(Menu menu) {
        super.onCreateOptionsMenu(menu);
        MenuInflater inf = getMenuInflater();
        inf.inflate(R.menu.menu, menu);
//      menu.findItem(R.id.main_menu_new).setIntent(
//                  new Intent(this, NewRecipe.class));
//      menu.findItem(R.id.main_menu_search).setIntent(
//                  new Intent(this, SearchRecipe.class));
//      menu.findItem(R.id.main_menu_options).setIntent(
//                  new Intent(this, Options.class));
              return true;
        }
```

continues

```
import android.view.Menu;
import android.view.MenuInflater;
import android.view.MenuItem;
```

This code instructs MainMenu to override the default menu creation with your new menu. The XML definition is pulled from **getMenuInflator()** and translated into a view. All returned menu items inherit the event **onOptionsItemSelected()**. The previous lines which are commented depend on classes which have not been implemented. Eventually, you will create activity classes for each menu item. The commented code is ready for use, but set to the side until the other pieces are in place.

Define the **onOptionsItemSelected()** method:

```
@Override
public boolean onOptionsItemSelected(MenuItem itm) {
super.onOptionsItemSelected(itm);
startActivity(itm.getIntent());
return true;
}
```

Figure 4.11: *The finished menu.*

Clicking the menu button shows and hides the menu just as you would expect, but there is a bug here. The logic in **onOptionsItemSelected()** assumes that a valid intent will be present. Clicking any of the menu items generates a fatal error, because the activity classes for the menu options have not been created yet. Modify this method to read:

```
@Override
public boolean onOptionsItemSelected(MenuItem itm) {
    super.onOptionsItemSelected(itm);
    Intent menuIntent = itm.getIntent();
    if (menuIntent != null)
    startActivity(menuIntent);
    return true;
}
```

Now, if the menu item is complete or returns null, startActivity() is not executed and a fatal error is averted.

PITFALL

Android does not always fail gracefully, especially with respect to handling null values. As a general rule, anticipate the null and much time debugging will be saved.

Storing Some Settings

As with menus, Android provides another simple way to implement one of the most basic features of any app: options. If your users are bilingual, you might want a preference for default language. Imagine that some users frequently request the app open to create a new recipe screen, but other users request that the app open to search by default. A preference for default activity solves the issue.

Creating a preference screen follows the same process as creating a standard activity:

ANDROID DOES

Eclipse allows copy/paste on classes. Select a .java class and **Edit** > **Copy** and **Edit** > **Paste**. Eclipse will even fix the name references inside the new class to match the new name you provide.

1. Create an activity in the Android Manifest named **.Options.**

2. Create a new Activity class in Package Explorer named **Options.**

3. Options.java should contain the following:

```
package com.recipesapp.basic;
import android.os.Bundle;
import android.preference.PreferenceActivity;
public class Options extends PreferenceActivity {
    @Override
    public void onCreate(Bundle savedInstanceState) {
        super.onCreate(savedInstanceState);
        addPreferencesFromResource(R.xml.options);
    }
}
```

Instead of extending the standard Activity class, this extends PreferenceActivity which contains specialized behavior for preferences.

4. Create an options.xml file of resource type Preference in the res/xml folder:

```
<?xml version="1.0" encoding="utf-8"?>
<PreferenceScreen xmlns:android="http://schemas.android.com/apk/res/
➥android">
        <PreferenceCategory android:title="About You">
                <EditTextPreference
                android:title="@string/opt_name_title"
                android:summary="@string/opt_name_summary"
                android:order="1"
                android:key="text"
                >
                </EditTextPreference>
        </PreferenceCategory>
        <PreferenceCategory android:title="Search Settings">
                <CheckBoxPreference
                android:title="@string/opt_search_title"
                android:summary="@string/opt_search_summary"
                android:defaultValue="true"
                android:order="2"
                android:key="check"
                >
                </CheckBoxPreference>
```

```
        </PreferenceCategory>
        <PreferenceCategory android:title="Personal Tastes">
                <ListPreference
                android:title="@string/opt_cuisine_title"
                android:summary="@string/opt_cuisine_summary"
                android:entries="@array/opt_cuisine_list"
                android:entryValues="@array/opt_cuisine_list"
                android:order="3"
                android:key="list"
                >
                </ListPreference>
        </PreferenceCategory>
</PreferenceScreen>
```

Preference categories are an optional way to nest and organize settings within the hierarchy. Android supports four basic types of preferences:

- **CheckBoxPreference.** The check box stores a true/false Boolean value. This preference is updated without a pop-up window.

- **ListPreference.** Lists use string arrays to provide entry and entry values, which can be different. Lists appear in a pop-up window.

- **EditTextPreference.** Provides a free text field to enter any value. Text preferences appear in a pop-up window.

- **RingtonePreference.** Provides an in app ringtone selection list.

Preferences share some of the same characteristics of other views, but Android will automatically manage storing and retrieving values. Define the strings from the previous XML in res/value/strings.xml:

```
        <string name="opt_cuisine_title">Default Cuisine</string>
        <string name="opt_cuisine_summary">Select the default cuisine
➥for search and new recipes</string>
        <string name="opt_search_title">Search</string>
        <string name="opt_search_summary">Search ingredients by
➥default</string>
        <string name="opt_name_title">Your Nickname</string>
        <string name="opt_name_summary">Enter a nickname to describe
➥yourself</string>
        <string-array name="opt_cuisine_list">
                <item>American</item>
                <item>Thai</item>
```

continues

```
        <item>Mediterranean</item>
        <item>Fusion</item>
        <item>Other</item>
    </string-array>
```

Everything is in place but a way to launch this new activity. If you recall from creating the menu, you implemented some code to do this, which is waiting behind comments. In MainMenu.java, uncomment the lines:

```
menu.findItem(R.id.main_menu_options).setIntent(
        new Intent(this, Options.class));
```

Run the app with these changes, and now select **Menu > Options.** See Figure 4.12.

Figure 4.12: *The Preferences menu for your recipe app.*

Modify some settings and marvel as they persist after exiting the app and opening it again. Ideas for settings will probably emerge either from user feedback or uncertainty about how to make your app function. In Chapters 6 and 15 you learn how to access stored data, including preferences, to use within your app.

Quitting Time

Few mobile apps actually have quit or exit buttons. It is usually implicit that by clicking the home or back keys the app will pause, effectively closing or switching to background mode. As a point in fact, the Android SDK does not even list a quit method. The Android OS handles app termination for you, so you can simply delete the Quit button.

The Least You Need to Know

- Eclipse handles renaming and reorganization through a process called refactoring.
- Styles can be applied to individual views only.
- Themes apply to the whole app or an entire activity.
- Menus provide quick access to core functions and are implemented similarly to other types of resources.
- Preferences are a special kind of resource that Android manages, mostly automatically, to store and retrieve defined settings.

Programming for Android

In This Chapter

- Core concepts in Java
- Android app life cycle
- Activities and views in depth
- Intents
- Introduction to BroadcastReceivers

The first few chapters have focused heavily on working with Java and XML in Eclipse to create some basic but fundamental extensions to an app. The Simply Recipes app now renders a default Welcome screen with a menu, an About page, an options menu, and a user preference page. You have applied styles and themes to customize the look and feel of the app.

Most of the core components of Android development have been referenced, at least in passing. This chapter focuses on Android development in more generic terms. Specifically, this chapter reviews the relationship between Java and the Android SDK.

What Is Java?

You already know that programs for Android are written using a combination of Java and XML. So what is Java and how does it fit into the Android OS? Java is a programming language like C or C++ that was developed by Sun Microsystems in the early 1990s. Sun, now Oracle, designed Java to enable developers to write applications which could run on any platform or OS with little or no modification.

ANDROID DOES

In 2009, Google released the Native Development Kit (NDK) for Android developers, enabling development in C and C++. Other tools exist that allow cross compilation from C++ to Java for Android.

Java is an object oriented language designed to be simple, robust, and familiar. Java has a simpler object model than C++, but uses much of its structure and syntax. This makes Java an easy transition from other languages. Android implements Java in a way that makes it even more simple and streamlined.

Object Oriented Programming Review

While developing for Android is straightforward in many ways, it does require a basic understanding of some of the core terminology. Please review some of the following terms:

- **Class.** A blueprint or template for creating new objects. Apple, as a class, would define the characteristics of all apples.

- **Object.** An instance of a class. Apple, as an object, would be the individual apple in your shopping cart.

- **Subclass.** A class that inherits definitional information from a parent class and adds additional information. McIntosh apples share the characteristics of all apples but have additional, distinguishing features.

- **Method.** A function that the class can perform. Methods require parameters and can return data of a certain type.

- **Property.** A variable. Equivalent to a method, except that it does not require parameters.

- **Inheritance.** The transfer of definitional data from a parent to a child class. Inheritance creates a hierarchical or cascading relationship from parent to children.

In your progress as an Android developer, you will encounter many other terms—polymorphism, abstraction, and decoupling—among others. For the purpose of this guide, the previous list should meet your needs.

The Java Virtual Machine

Java's capability to run the same code on many, vastly different operating systems is made possible by the *Java Virtual Machine* (*JVM*). You may have already noticed that Eclipse is a Java application running in a JVM.

> **DEFINITION**
>
> The **Java Virtual Machine (JVM)** is an application written for a specific platform, which translates platform-neutral Java code into code that the host OS can run.

Oracle provides JVMs for most operating systems, including Windows, Mac, and Linux. In order for the Java you write in Eclipse to run on Android, it must run in a virtual machine (VM). Android implements the Dalvik VM, which shares some similarities to Oracle's JVMs. Most notably, the Java code you write for Android will always run, because all versions of Android implement the Dalvik VM.

As the Dalvik VM is not the same as Oracle's JVM, there are a number of differences. Some Java code will not work with Android. Key factors to consider:

- **Performance.** Dalvik was designed for mobile devices, optimized for devices with little memory. Apps that assume the availability of significant memory or processing power may not perform well.

- **Shared code.** Android allows you to incorporate any third-party Java code that is compatible with Dalvik. However, just because the Java code you found builds successfully under Eclipse does not mean that it was written with Android devices in mind.

- **Java version.** Android OS updates are frequent, but so are updates to Java. Not all features of the latest Oracle JVM or changes to Java *APIs* will work with Android.

> **DEFINITION**
>
> Many programming languages will provide an **Application Programming Interface (API).** An API is a guide to writing code in a language. The API provides quick exposure to available elements of a language.

All Android apps run in their own instance of a Dalvik VM. This allows Android to sandbox apps, a security and stability consideration for the OS. Because each app runs isolated in a VM, malicious apps are less likely to destabilize the OS. It also prevents apps from negatively interfering with each other.

Keep in mind that Oracle JVM does one particular thing better than Dalvik VM: runtime bytecode generation, which allows direct access to Java libraries. It is not absolutely necessary and, for our needs, won't be an issue.

The Front of an Android Application

Take a look at your mobile device and think about the behavior you expect from it. If you are holding it and interacting with it, you are probably interested in executing a task or series of tasks. You might open a browser to check the stock market, send an e-mail, and start a game of solitaire, when you might receive a call. When you end the call, what screen are you now looking at? Is it call history in the phone app or is it the solitaire app?

At their core, apps are collections of activities. Much of the attention Android gives its apps focuses on moving between activities and their states.

Activities

Activities are actions within the app life cycle. Android manages every running activity object in an Activity stack. The Activity class defines a series of *methods* that instruct each Activity object how to behave in response to the basic user workflow, which involves starting, switching, and resuming activities. This workflow applies not only to all the activities in your app, but to all activities in all running apps.

Launching the browser app starts a new activity. Android looks at the top activity in the stack, perhaps from the home screen where you tapped the app icon, and pauses it. Android then places the browser app at the top of the stack. An incoming call at this point would pause the current browser activity and place the call activity at the top of the stack. Once the call activity finishes, Android removes it from the stack and resumes the browser activity.

The life cycle of an activity can be considered in three phases:

- **The entire lifetime.** This period lasts from the call to onCreate() through onDestroy().

- **The visible lifetime.** This begins at onStart() and persists until onStop(). During this time, the Activity can respond to user input, but it may not be in the foreground, at the top of the Activity stack.

- **The foreground lifetime.** Between onResume() and onPause(), the Activity is at the top of the stack and is the activity in focus. Your app will frequently transition between these two methods, for example every time the device goes to sleep and wakes up.

PITFALL

Reserve complex operations for onCreate() and onStart(), which are called only once and a few times respectively. onResume() is likely called many more times in the Activity life cycle. Frequently called methods should execute quickly.

Each Activity life cycle method has different attributes, including if they can be stopped by you, the programmer, once initiated. The following table provides an overview.

The Methods of the Activity Life Cycle

Lifetime	Method	Description	Killable?	Followed By
Entire	onCreate()	Called when an activity is first started. Your activity layout should normally be placed in this method. onCreate() takes a bundle parameter, which can be used to restore the activity state if the activity is unexpectedly terminated.	No	onRestart() or onStart()
Visible	onRestart()	Called after onStop() but before onStart() is called again. onStart() always follows.	No	onStart()
Visible	onStart()	Called as the activity becomes visible.	No	onResume() or onStop()
Foreground	onResume()	The activity has moved to the top of the Activity stack and is becoming interactive.	No	onPause()

continues

The Methods of the Activity Life Cycle   (continued)

Lifetime	Method	Description	Killable?	Followed By
Foreground	**onPause()**	Called before another activity is resumed. Ideal time to store unsaved changes.	Yes	onResume() or onStop()
Visible	**onStop()**	Called when a new activity is started, an existing activity is resumed, or if the active app is being destroyed.	Yes	onRestart() or onDestroy()
Entire	**onDestroy()**	The last method call your activity receives. If Android is terminating your activity, the isFinishing() method will return false.	Yes	Nothing

Until onPause() is called, the activity is front and center; however, as soon as onPause() is called, Android reserves the right to terminate your activity. The Dalvik VM will attempt to optimize each instance of every app for best performance, but sometimes Android simply runs out of resources and needs to select apps to kill. This means that onPause() is your last chance to preserve your user's information before going into the background. onPause() is also a great time to release any resources your app does not require for background operation.

Views

Where an app is a collection of activities, an activity is a hierarchy of views. Views are the building blocks of all Android applications. The View class itself is quite diverse. Its largest *subclasses* are ViewGroup and TextView. You have already worked with objects of the ViewGroup subclass, namely layouts like RelativeLayout and LinearLayout. You should also be familiar with some TextView children such as Button and CheckBox.

DEFINITION

A **subclass** inherits some of the methods and properties of a class. Subclasses share enough class characteristics to be considered related but are unique enough to deserve their own definition.

A user interface or activity is composed of hierarchy objects of either the View class or the ViewGroup class. All of these visual elements are related in a class hierarchy that begins with the View class. This means that objects of this class and subclasses share common base methods, shown in the following table.

Common View Methods

XML Attribute Name	Method	Description
android:background	setBackgroundResource(int)	Set the background a drawable resource.
android:clickable	setClickable(boolean)	True if this view responds to click events.
android:contentDescription	setContentDescription (CharSequence)	Description of the view's content.
android:duplicateParentState		True if the view gets its drawable state (focused, pressed, etc.) from its parent.
android:fadingEdge	setVerticalFadingEdgeEnabled (boolean)	Should be faded on. Defines which edges scrolling.
android:fadingEdgeLength	getVerticalFadingEdgeLength()	Defines the length of the fading edges.
android:focusable	setFocusable(boolean)	True if the view can take focus.
android:hapticFeedbackEnabled	setHapticFeedbackEnabled(boolean)	True if the view should have haptic feedback enabled for events such as long presses.

continues

Common View Methods (continued)

XML Attribute Name	Method	Description
android:id	setId(int)	Set an identifier to later reference the view by calling findViewById().
android:keepScreenOn	setKeepScreenOn(boolean)	Controls whether the view's window should keep the screen on while visible.
android:longClickable	setLongClickable(boolean)	True if this view reacts to long click events.
android:minHeight		The minimum height of the view.
android:minWidth		The minimum width of the view.
android:nextFocusDown	setNextFocusDownId(int)	Defines the next view to give focus to when the next focus is FOCUS_DOWN.
android:nextFocusLeft	setNextFocusLeftId(int)	Defines the next view to give focus to when the next focus is FOCUS_LEFT.
android:nextFocusRight	setNextFocusRightId(int)	Defines the next view to give focus to when the next focus is FOCUS_RIGHT.
android:nextFocusUp	setNextFocusUpId(int)	Defines the next view to give focus to when the next focus is FOCUS_UP.
android:onClick		Name of the method in this View's context to invoke when the view is clicked.

XML Attribute Name	Method	Description
android:padding	setPadding(int,int,int,int)	Sets the padding, in pixels, of all four edges.
android:paddingBottom	setPadding(int,int,int,int)	Sets the padding, in pixels, of the bottom edge; see padding.
android:paddingLeft	setPadding(int,int,int,int)	Sets the padding, in pixels, of the left edge; see padding.
android:paddingRight	setPadding(int,int,int,int)	Sets the padding, in pixels, of the right edge; see padding.
android:paddingTop	setPadding(int,int,int,int)	Sets the padding, in pixels, of the top edge; see padding.
android:saveEnabled	setSaveEnabled(boolean)	If unset, no state will be saved for this view when it is being frozen.

PITFALL

When using methods that reference other objects, as in setFocusDownId(), be sure the destination object exists. If the view does not exist or is not visible, a RuntimeException will result.

These methods are accessible to classes like TableLayout, because TableLayout extends the LinearLayout class, which extends the ViewGroup class, which extends the View class. Inheritance allows base methods to cascade to all children, even those many times removed. This cascade is parent to child only. The methods unique to the LinearLayout class cascade down to TableLayout but not up to ViewGroup.

GOOGLE IT

The Android Developer's Guide provides some helpful illustrations of various, frequently used View subclasses at http://developer.android.com/guide/tutorials/views/index.html.

Behind the Scenes

The user interface of your app is entirely defined by Activities and Views, which you might think of as the front end of the app. Consisting mostly of static resources including XML, Activities and Views do not do very much. Fortunately, parallel classes exist to service the back end.

Intents

Although some pseudo-procedural code can be implemented through XML, most Android actions require Intents. In Chapter 4, you implemented an *Intent class* to launch the About screen. Without intents, your activities have no way to translate actions into responses. With an intent, you can connect clicking a button with some action, like starting a new activity.

DEFINITION

The **Intent class** describes a kind of operation to perform. Intents can start an Activity, communicate to a BroadcastReceiver, start a Service, or interact with a Service.

When a user touches the About button in Simply Recipes, they probably do not care how the screen appears or what activities or intents might be. The user just cares about the result. Similarly, Android activities have an apathetic attitude toward intents. Intents are designed to decouple the interface from the operations being performed.

Intents are most frequently used to launch activities. To function, an intent needs some information:

- **Action.** Some action to perform, usually a constant such as ACTION_VIEW, ACTION_EDIT, or ACTION_MAIN.

- **Data.** Some data to operate on, such as a contact. This must be expressed as a *Uniform Resource Identifier*, or URI.

- **Category.** Gives additional information about the action to execute.

- **Type.** Specifies an explicit MIME type for the intent data. Type is inferred from the data itself unless this attribute is set.

- **Component.** Explicitly defines the component class to use for the intent. This is normally inferred from action and type.

- **Extras.** A bundle of any additional information. This can contain any extra data relevant to the event.

DEFINITION

A **Uniform Resource Identifier (URI)** is an address for the location of some information. On the web, this could be a web address like http://www.google.com. In Android, this is likely to be a local address like content://contacts/people/1.

Combining these attributes together, you are able to send complete commands like, "Dial Joe's mobile phone number" or "Open Maps app to my current location." Android offers two paths to defining these commands: explicit intents and implicit intents.

Explicit intents specify an exact component and class to run. These can be useful for launching activities within your app. You will notice that you used an explicit intent in Chapter 4 to launch the About screen:

```
Intent showAbout = new Intent(this, About.class);
startActivity(showAbout);
```

Implicit intents allow Android to locate the right component to run the intent. Action, Data, and Category are used to query the PackageManager for the most appropriate component. To implicitly call a contact, for example, you would use:

```
Intent i = new Intent(Intent.ACTION_CALL,
Uri.parse("content://contacts/people/1"));
startActivity(i);
```

Common Intent Actions

Action	Description
ACTION_EDIT content://contacts/people/1	Open contact with identifier "1" to edit.
ACTION_VIEW tel:123	Open phone app with "123" in the dialer.
ACTION_VIEW content://contacts/people/1	Open contact "1" to view.
ACTION_CALL tel:123	Start call with phone number "123."
ACTION_DIAL content://contacts/people/1	Open phone app with contact's number, but do not call.

Notice that supplying a telephone number to ACTION_VIEW launches the phone app, while supplying a contact opens the contact app. In the absence of more data, such as a category, Android will make a best guess for which activity to launch based on the kind of data supplied.

Intent Filters

While most of your intents will be wired into your activities used to move back and forth between the different screens of your app, there is another way to use intents. If an activity must be called by an intent, what launches the default activity of your app when you open it? The intent filter in your AndroidManifest.xml:

```
<activity android:label="@string/app_name" android:name=".
MainMenu">
        <intent-filter>
            <action android:name="android.intent.action.MAIN" />
            <category android:name="android.intent.category.
LAUNCHER" />
        </intent-filter>
    </activity>
```

The action android.intent.action.MAIN declares that .MainMenu is the default activity, and android.intent.category.LAUNCHER instructs the activity to open by default. As the name suggests, intent filters are used to limit the number of matching implicit intent actions that your activity's intents might use.

```
<intent-filter>
<action android:name="android.intent.action.VIEW" />
<category android:name="android.intent.category.BROWSEABLE" />
<category android:name="android.intent.category.DEFAULT" />
</intent-filter>
```

This intent filter when applied to an activity instructs that VIEW actions should only query the PackageManager for tasks that fit into the BROWSEABLE and DEFAULT categories.

You can also allow your activity to be accessible to implicit events called from other apps by adding the ALTERNATIVE and SELECTED_ALTERNATIVE categories to an intent filter:

```
<category android:name="android.intent.category.ALTERNATIVE" />
<category android:name="android.intent.category.SELECTED_ALTERNATIVE"
 />
```

Background Receivers

While you will most frequently use intents to interact with activities, they do serve two other useful functions. They can broadcast events and work with services. *BroadcastReceivers* are a special kind of background task, dedicated to listening for event broadcasts and responding with an action.

> **DEFINITION**
>
> **BroadcastReceiver** class members listen for broadcasts sent by intents and can send notifications, run intents, start services, or update the display. BroadcastReceivers are started when a matching intent broadcast is sent.

No need to implement the following code, but let's suppose our recipe app had a spice inventory function, and the user begins entering a recipe which includes a spice they do not have. You could fire an intent broadcast which includes the name of the new spice, and a BroadcastReceiver listening for new spices could add this to a shopping list.

Creating an intent broadcast is straightforward:

```
        public static final String NEW_INGREDIENT = "com.recipesapp.
basic.action.NEW_INGREDIENT";
        Intent missingIngredient = new Intent(NEW_INGREDIENT);
        missingIngredient.putStringExtra("spice","paprika");
sendBroadcast(missingIngredient);
Based on matching intent filters, a matching BroadcastReceiver must
exist in order for the desired action to happen.
        import android.content.BroadcastReceiver;
        import android.content.Context;
        import android.content.Intent;
        public class NewIngredientDetectedToAdd extends
BroadcastReceiver {
@Override
public void onReceive(Context context, Intent intentt) {
Uri data = intent.getData();
String spice = data.getStringExtra("spice");
//Now do something
context.StartActivity(..);
}
};
```

The onReceive() method must complete within five seconds, or it will fail. Unlike services, BackgroundReceivers do not run constantly in the background. Rather, Android dynamically starts them as needed by intent broadcasts.

The BroadcastReceiver must be registered, just like Activities. This can be done in AndroidManifest.xml:

```
<receiver android:name=".NewIngredientDetectedToAdd">
<intent-filter>
<action android:name="com.recipesapp.basic.action.NEW_INGREDIENT" />
</intent-filter>
</receiver>
```

Registering a BroadcastReceiver in the manifest ensures that it always receives broadcasts, even if your app is not yet started. Sometimes, you want your BroadcastReceivers only to be accessible within your app or while your app is running. In this case, you can register the receiver in code:

```
    IntentFilter filter = newIntentFilter(NEW_INGREDIENT);
    NewIngredientDetectedToAdd receiver = new
➥NewIngredientDetectedToAdd();
    registerReceiver(receiver, filter);
Later in your code, when you have finished with the receiver, simply
➥unregister it:
    unregisterReceiver(receiver);
```

Understanding the anatomy of your Android app can assist you enormously as you begin to improve the user interface and extend the functionality of the app. Intents and BroadcastReceivers provide easy and powerful ways to execute the core actions of your app. In Chapter 14, you learn more about background processes and services.

In the next chapter, you return to expanding your practical knowledge of Android by implementing 2D graphics and beginning to collect user input.

The Least You Need to Know

- All Android apps run in independent Dalvik virtual machines.
- Most, but not all, features of the Java language work with Android.
- The Activity life cycle has three phases: entire, visible, and foreground.
- Each Activity is composed of a View hierarchy.
- Intents are the primary mechanism to switch between Activities.

Constructing Your Application

Refining your user interface will be one of the most important parts of your app development. Building interfaces in XML is surprisingly straightforward on Android. This part will take you through extending your app across multiple screens, collecting and working with user input, and rendering graphics.

Resources and Animation

In This Chapter

- Creating a splash screen
- Working with animation resources
- Creating animation effects
- Using drawables to render content

If you have opened many apps on your mobile device, you might have noticed that some immediately open and others display an intermediate splash screen. Splash screens offer a chance to execute an initialization task in the background or to brand your app with a unique look and feel. They also happen to be a great way to begin working with images and two-dimensional animation.

By this point, you are familiar with creating new Activity classes and Views, and adding and modifying resources. In this chapter, you add some incremental knowledge to your developer's toolkit. This chapter introduces some new ways to use resources and animations, and you build a functional splash screen for Simply Recipes.

The Splash View

Animation attracts. It is an easy way to add polish to your application and an effective way to communicate different kinds of information to your users. An animated progress bar indicates that some task runs but has not completed. Animations are great transitions, showing that one activity has ended and another has started.

The purpose of a splash screen is to render some simple text or images for a few seconds before opening the main application. Functionally, the splash screen is no different than any other screen in your Android app. It is unique only in the way that you design it inside the app. You can just as easily use the About screen as a splash screen, but it would lack some visual appeal.

Because you are creating just another Activity, you probably already have a sense for how the basic strategy looks:

- Add a new **Splash** Activity to AndroidManifest.xml.
- Create a new **Splash** Activity class.
- Create a new **Splash** XML resource file for both the layout and layout-land folder.

After you have finished these steps using the BasicRecipes project in Eclipse, you can get started. Soon you will add new customized resources for colors, dimensions, and animations.

Assemble the View

As your Activities become more sophisticated, it becomes useful to start sketching the View hierarchy of your screen. Like writing an outline, this can help you visualize the work ahead and serve as a reference when you start troubleshooting.

This Activity is relatively straightforward: a single LinearLayout with two TextViews and an ImageView. Thinking further ahead, you can imagine what animation effects you want to apply to the elements in this ViewGroup. It is sketched out in Figure 6.1.

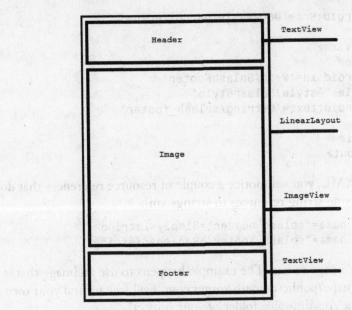

Figure 6.1: *The rough sketch of the app.*

With a plan in place, you can now edit splash.xml to create the Splash Activity view:

```xml
<?xml version="1.0" encoding="utf-8"?>
<LinearLayout
    xmlns:android="http://schemas.android.com/apk/res/android"
android:id="@+id/SplashLayoutRoot"
    android:orientation="vertical"
    android:layout_width="fill_parent"
    android:layout_height="fill_parent"
    android:layout_gravity="center"
    android:gravity="center"
    android:background="@color/splash_background">
    <TextView
        android:id="@+id/SplashHeader"
        style="@style/SplashStyle"
        android:text="@string/splash_header"
        >
    </TextView>
    <ImageView
        android:id="@+id/SplashImageCenter"
        android:layout_width="wrap_content"
        android:layout_height="wrap_content"
        android:layout_gravity="center"
```

continues

```
        android:src="@drawable/parrotrecipe"
        >
    </ImageView>
    <TextView
        android:id="@+id/SplashFooter"
        style="@style/SplashStyle"
        android:text="@string/splash_footer"
        >
    </TextView>
</LinearLayout>
```

Looking at the XML, you will notice a couple of resource references that do not exist yet. Add the missing string resources to strings.xml:

```
<string name="splash_header">Simply</string>
<string name="splash_footer">Recipes</string>
```

You also need an image to use. The example happens to use an image that is 234 × 324 pixels from http://public-domain.zorger.com. Feel free to find your own image and put it into the /res/drawable folder of your project.

GOOGLE IT

Public Domain images are guaranteed to be free for your commercial and non-commercial use. The images used in this text come from the Public Domain unless otherwise noted.

With string and drawable resources in place, color and style resources remain to be located.

Colors and Dimensions

Many of the classes necessary to accessing drawing functionality are located in the android.graphics package, which includes canvases, color filters, rectangles, and points. This also includes *color*. Quite limited compared to other classes, the Color class provides a simple way to represent colors as integer values. Android provides a few ways to access these:

DEFINITION

A **color** in Android is the hexadecimal combination of alpha, red, green, and blue. Alpha represents transparency. Opaque black would be ff000000 while transparent white would be 00ffffff.

- **By Name.** Some common colors, just as in HTML, are accessible by name. This includes black, blue, red, white, yellow, green, and a few others. In Java, you could access these colors with this syntax:

```
int magenta = Color.MAGENTA;
```

- **By Value.** If you know the red, green, and blue integer values, you can use a static factory on the Color class. In Java:

```
magenta = Color.argb(255, 255, 0, 255);
```

- **By Hexadecimal.** Perhaps the easiest way to access colors is to store the hex value in an XML resource for easy access by name. Inside a resources file:

```
<color name="dark_magenta">#ff8B008B</color>
```

Anywhere else in your XML resources, the color is now accessible by name:

```
android:textColor="@color/dark_magenta"
```

The same color is also accessible in Java through your R class:

```
magenta = getResources().getColor(R.color.dark_magenta);
```

To get started with the Splash activity, create a new colors.xml in the layout folder. It will have a light blue for splash_header_text and a dark gray for splash_background:

```
<?xml version="1.0" encoding="utf-8"?>
<resources>
    <color name="splash_header_text">#ff3b9c9d</color>
    <color name="splash_background">#ff2b2d33</color>
</resources>
```

You now have some colors defined for the splash screen. It is time to add some text dimensions. Create a new dimens.xml in the layout folder with some size definitions:

```
<?xml version="1.0" encoding="utf-8"?>
<resources>
    <dimen name="splash_text_size">40dp</dimen>
    <dimen name="splash_text_spacing">2dp</dimen>
</resources>
```

You can put these to immediate use in a style for this activity. Using the styles.xml resource you created in Chapter 4, add a new SplashStyle style to your project (see Figure 6.2):

```
<style name="SplashStyle" parent="@android:style/TextAppearance.
  ↪Medium">
    <item name="android:layout_width">fill_parent</item>
    <item name="android:layout_height">wrap_content</item>
    <item name="android:layout_gravity">center</item>
    <item name="android:gravity">center</item>
    <item name="android:textSize">@dimen/splash_text_size</item>
    <item name="android:textColor">@color/splash_header_text</
  ↪item>
    <item name="android:lineSpacingExtra">@dimen/splash_text_
  ↪spacing</item>
</style>
```

Figure 6.2: *The splash image in the workspace.*

The splash.xml resource now points to valid resources and should correctly render inside Eclipse.

Ready, Set, Animate

If you think about creating animation in a pen and paper sense, there are really only two ways to do it. You can draw a series of events or you could take an existing drawing and manipulate it. Android treats animation the same way—either you define a drawing to render on a canvas, or you use an existing drawable resource and add some effects to it.

Using your existing /res/drawable/ resource, you can add some *tween* effects. Animation resources live in the /res/anim folder. Tween transitions include:

- **Alpha.** Takes an object from one level of transparency to another. An alpha value of 1.0 indicates complete opacity, while a value of 0.0 indicates complete transparency.

- **Scale.** Changes the size of an object on an X/Y scale. X/Y Scale values indicate the size offset where 1.0 indicates no change.

- **Translate.** Moves an object vertically or horizontally.

- **Rotate.** Spins an object from one angular position to another on a defined axis.

- **Set.** A collection of transition effects.

DEFINITION

Tween animations are defined in XML and describe transition effects to visually move objects from one state to another over some length of time.

For this splash screen, create animation resources to apply to the header and footer text as splash_text.xml:

```xml
<?xml version="1.0" encoding="utf-8" ?>
<set
    xmlns:android="http://schemas.android.com/apk/res/android"
    android:shareInterpolator="false">
    <alpha
        android:fromAlpha="0.0"
        android:toAlpha="1.0"
        android:duration="2500"
        >
    </alpha>
</set>
```

This tween animation takes some object from complete transparency to opacity over the course of 2.5 seconds, or 2500 milliseconds. This type of animation also supports the use of *interpolators*, which customize the way an animation progresses. For example, accelerateInterpolator instructs an animation to begin slowly and then gradually accelerate.

DEFINITION

Interpolators modify animation effects by increasing or decreasing the speed, bouncing or repeating the effects.

With a text effect in place for later use, now create a animation resource for the center image, **splash_image.xml:**

```xml
<?xml version="1.0" encoding="utf-8"?>
<set xmlns:android="http://schemas.android.com/apk/res/android"
        android:shareInterpolator="false">
        <alpha
android:interpolator="@android:anim/decelerate_interpolator"
        android:fromAlpha="0.2"
        android:toAlpha="1.0"
        android:duration="3000">
        <rotate
                android:fromDegrees="0"
                android:toDegrees="360"
                android:pivotX="45%"
                android:pivotY="55%"
                android:startOffset="500"
                android:duration="2500"
        />
        <scale
                android:interpolator="@android:anim/accelerate_
    interpolator"
                android:fromXScale=".1"
                android:fromYScale=".1"
                android:toXScale="1.0"
                android:toYScale="1.0"
                android:pivotX="50%"
                android:pivotY="50%"
                android:duration="3000"
                />
        </alpha>
</set>
```

The animation for this resource is slightly more sophisticated. First, notice that the types of animations can be nested, just as other View classes can be nested. All transition effects are applied at the same time, unless a startOffset is defined, which can be done at any level.

Method Acting

With all of the requisite resources now in place, you can implement the logic for these animations by writing the necessary Java code for the Splash Activity. It is the job of the Splash class to connect the various resources you have created and execute some actions upon them. This requires accessing packages outside the Activity class, so begin by adding the necessary imports:

```
package com.recipesapp.basic;
import android.app.Activity;
import android.content.Intent;
import android.os.Bundle;
import android.view.animation.Animation;
import android.view.animation.AnimationUtils;
import android.view.animation.Animation.AnimationListener;
import android.widget.TextView;
import android.widget.ImageView;
```

ANDROID DOES

Sometimes it can be difficult to know which imports to use in your class. Fortunately, if you miss one, Eclipse detects the error and prompts you to add the import. When uncertain, be sure to watch your Problems view in Eclipse.

With the correct references in place, it is time to implement the class and at least one life cycle method, onCreate():

```
public class Splash extends Activity {

    @Override
    public void onCreate(Bundle savedInstanceState) {
        super.onCreate(savedInstanceState);
        setContentView(R.layout.splash);
        beginAnimation();
    }
}
```

There is very little new here. The class extends Activity, sets the content view to the splash.xml resource, and then calls the beginAnimation() method which you can now create:

```
private void beginAnimation() {
// Header animation
TextView header = (TextView) findViewById(R.id.SplashHeader);
Animation headerAnim = AnimationUtils.loadAnimation(this, R.anim.
   ➡splash_text);
header.startAnimation(headerAnim);
}
```

These three lines provide the magic needed to animate a View object. The first steps instance a TextView header as the SplashHeader View object. Next, the splash_text.xml animation resource is assigned to an Animation property headerAnim. Finally, you instruct Android to startAnimation() on header using headerAnim. Using this syntax, create animations for the remaining two View objects in the Splash layout.

So far so good, but the splash screen needs to do one more crucial task: redirect to the Welcome screen when the animations have finished. Fortunately, Android makes this simple by providing animation listeners:

```
// Prepare end of animation event
imageAnim.setAnimationListener(new AnimationListener() {
public void onAnimationEnd(Animation animation) {
// All done, open the Main Menu
startActivity(new Intent(Splash.this, MainMenu.class));
}
//Required method, nothing to do here
public void onAnimationRepeat(Animation animation) {
}
//Required method, nothing to do here
public void onAnimationStart(Animation animation) {
}
});
```

Depending on how you have created your animations, some may last longer than others. In this case, the image animation lasts 3 seconds while the text animation lasts 2.5 seconds. To detect when the animation is over, the longest running animation must be selected. Replace imageAnim with the name you selected.

The setAnimationLister() method inherits three required methods: onAnimationEnd(), onAnimationRepeat(), and onAnimationRestart(). For this task, only onAnimationEnd() matters, and the only required action is to launch MainMenu.

While 3 seconds is not a long time, it is enough time for something else to happen. Perhaps the user receives a call in the middle of the animation or perhaps the user hits the home screen button. What if the animation is interrupted? Fortunately, the Activity life cycle as discussed in Chapter 5 provides an answer: the onPause() and onResume() methods.

To ensure that the user always receives the full splash screen animation from start to finish, you can flesh out these life cycle methods to leave your splash screen in a clean state. First, define onPause():

```
@Override
protected void onPause() {
    super.onPause();

    // Clear the animation. We'll start fresh on resume.
    TextView header = (TextView) findViewById(R.id.SplashHeader);
    header.clearAnimation();
}
```

Don't forget to include all the View objects you used in onCreate(). Next, define an action for onResume(). Because the state of all animations is now cleared, you can simply call beginAnimation() again:

```
@Override
protected void onResume() {
    super.onResume();

    //Start the animation from the beginning.
    beginAnimation();
}
```

Your completed class should now look something like the following:

```
package com.recipesapp.basic;
import android.app.Activity;
import android.content.Intent;
import android.os.Bundle;
import android.view.animation.Animation;
import android.view.animation.AnimationUtils;
import android.view.animation.Animation.AnimationListener;
import android.widget.TextView;
import android.widget.ImageView;
public class Splash extends Activity {

    @Override
    public void onCreate(Bundle savedInstanceState) {
```

continues

```java
        super.onCreate(savedInstanceState);
        setContentView(R.layout.splash);
        beginAnimation();
    }
    private void beginAnimation() {

        // Header animation
        TextView header = (TextView) findViewById(R.id.SplashHeader);
        Animation headerAnim = AnimationUtils.loadAnimation(this,
R.anim.splash_text);
        header.startAnimation(headerAnim);

        // Footer animation
        TextView footer = (TextView) findViewById(R.id.SplashFooter);
        Animation footerAnim = AnimationUtils.loadAnimation(this,
R.anim.splash_text);
        footer.startAnimation(footerAnim);

        // Image Animation
        ImageView image = (ImageView) findViewById(R.
id.SplashImageCenter);
        Animation imageAnim = AnimationUtils.loadAnimation(this,
R.anim.splash_image);
        image.startAnimation(imageAnim);

        // Prepare end of animation event
        imageAnim.setAnimationListener(new AnimationListener() {
            public void onAnimationEnd(Animation animation) {

            // All done, open the Main Menu
                startActivity(new Intent(Splash.this, MainMenu.
class));
            }
            //Required method, nothing to do here
            public void onAnimationRepeat(Animation animation) {
            }
            //Required method, nothing to do here
            public void onAnimationStart(Animation animation) {
            }
        });
    }
    @Override
    protected void onPause() {
        super.onPause();

        // Clear the animation. Start fresh on resume.
```

```
        TextView header = (TextView) findViewById(R.id.SplashHeader);
        header.clearAnimation();

        TextView footer = (TextView) findViewById(R.id.SplashFooter);
        footer.clearAnimation();

        ImageView image = (ImageView) findViewById(R.
    id.SplashImageCenter);
        image.clearAnimation();
    }
    @Override
    protected void onResume() {
        super.onResume();

        //Start the animation from the beginning.
        beginAnimation();
    }
}
```

Remember to update the Intent Filter in AndroidManifest.xml to use .Splash as the launched activity. Once done, you should be able to run Simply Recipes in the emulator to see a gradual fade of the header and footer, with a fade, spin, and scale of the center image (see Figure 6.3).

Figure 6.3: *The final splash screen for the recipe app.*

2D Graphics Overview

While any given screen in Android might look like a fairly static document or image, the contents of the View for the current Activity must be drawn on creation. Up to now, this depended on layout resources to instruct the Activity on what content to place where.

View elements can also be constructed in code. This makes it possible to create dynamic visual content. With two-dimensional images, Android allows animation through one of two basic paths: using an existing image, or *drawable*, or by drawing new graphics. Drawables fit into three categories:

- **Bitmap.** Supported file types include PNG (preferred), JPEG (acceptable), and GIF (discouraged).

- **Shape.** A configurable shape which can be drawn based, such as an oval, circle, square, or rectangle.

- **Nine Patch.** A special type of PNG image that includes nine stretchable areas, which allow Android to dynamically resize the image based on screen size. The default button images are examples of NinePatchDrawables.

DEFINITION

A **drawable** is any resource which can be drawn or rendered as a graphic on the screen. The Drawable class defines a few generic methods to interact with the object being drawn.

In addition to tweening and the methods available on these types of drawables, Android also supports one other way to animate images with frame animation. Frame animation is implemented identically to tweening, but the animation resource is defined differently:

```
<animation-list xmlns:android="http://schemas.android.com/apk/res/
  android"
   android:oneshot="true">
   <item android:drawable="@drawable/boiling_pot1"
  android:duration="100" />
   <item android:drawable="@drawable/boiling_pot2"
  android:duration="100" />
   <item android:drawable="@drawable/boiling_pot3"
  android:duration="100" />
</animation-list>
```

This instructs Android to load each image sequentially for the specified duration.

Working with Drawables

You can imagine a slideshow app that displays a series of images. You could implement this as a frame animation. But what if the user wanted to pause the animation or go back and forth between images? This would be difficult to do entirely in XML, but Android makes this easy to do in Java.

Taking a drawable resource, you can actually change a View object from code. Suppose you had an ImageView that needed some input from the user to determine which graphic to display. This could be a user clicking a next or back button, or it could be a life cycle event on the Activity. If you wanted to create the Splash image and view in code instead of in XML, you might create a LinearLayout and ImageView at onCreate() to set the image:

```
LinearLayout splashLayout;
protected void onCreate(Bundle savedInstanceState) {
    super.onCreate(savedInstanceState);
// instance a Splash Linear Layout
splashLayout= new LinearLayout(this);
// instance an ImageView
ImageView splashImage = new ImageView(this);
splashImage.setImageResource(R.drawable.parrotrecipe);
// set the splashImage boundary equal to the image's dimensions
splashImage.setAdjustViewBounds(true);
splashImage.setLayoutParams(new
Gallery.LayoutParams(LayoutParams.WRAP_CONTENT, LayoutParams.WRAP_
    ➥CONTENT));
// add splashLayout to the layout
splashLayout.addView(splashImage);
setContentView(splashLayout);
}
```

Using Shapes

No need to implement it in the recipe app, but in addition to working with and modifying a layout of an Activity, drawables can be instanced and manipulated as well. This could be to scale or stretch a bitmap or to draw a new shape. Just as you can define layout resources in XML, you can also define them in code:

```
public class OvalView extends View {
private ShapeDrawable oval;

public OvalView (Context context) {
```

continues

```
        super(context);

        int left = 10;
        int top = 10;
        int width = 100;
        int height = 300;
        int right = left + width;
        int bottom = top + height;
        oval = new ShapeDrawable(new OvalShape());
        oval.getPaint().setColor(Color.CYAN);
        oval.setBounds(left, top, right, bottom);
    }
    protected void onDraw(Canvas canvas) {
        oval.draw(canvas);
    }
}
```

Rather than referencing a layout resource, the OvalView class provides the content:

```
protected void onCreate(Bundle savedInstanceState) {
super.onCreate(savedInstanceState);
    setContentView(new OvalView(this));
}
```

An instance of this OvalView class will render a tall, cyan oval. While it is not particularly thrilling to behold at the moment, it hints at the possibilities available to you by drawing your own shapes. Part of this process involves a few new classes worth highlighting:

- **Paint.** This class contains the style and color information needed to draw most graphics.

- **Canvas.** Like real world canvases, a Canvas begins as a blank slate. The View class contains an onDraw() method which provides the opportunity to draw your own content.

- **Path.** Vector shapes are stored in this class, such as lines and rectangles. These can be manipulated to form other shapes.

Using vectors, you can draw a similar oval:

```
public class Oval extends View {
        private Path oval;
        private RectF bounds;
        private Paint cyan;
```

```
public Oval (Context context) {
        super(context);
        bounds = new RectF();
        bounds.bottom = 320;
        bounds.left = 20;
        bounds.right = 120;
        bounds.top = 20;
        cyan = new Paint();
        cyan.setColor(Color.CYAN);
        oval = new Path();
        oval.addOval(bounds, Direction.CCW);
        }
protected void onDraw(Canvas canvas) {
canvas.drawPath(oval, cyan);
        }
}
```

The syntax is similar, but Paths allow you to create more flexible shapes, which you can manipulate by adding text or modifying the way the lines are drawn.

Your decisions for how to implement a graphic or drawable will vary by the needs of your Activity. For images that change only infrequently, you will likely use drawable resources and XML definitions. If you were to implement a game of tic-tac-toe, you would likely use vector graphics to render the screen.

Now that you have implemented some animation, you can look ahead to building more practical functionality into your app. In the next chapter, you learn how to build the New Recipe screen using text and input widgets.

The Least You Need to Know

- Tween animation applies transitional effects on static images.
- Animation resources live in the /res/anim folder.
- Drawables represent any graphic which can be drawn in a View.
- PNG is the preferred drawable resource format, but JPEG and GIF will work.
- Shapes and Paths provide a way to draw free lines or basic shapes on a Canvas.

Building Input and Output

In This Chapter

- Building complex views
- Using widgets for user input
- Working with information in lists
- Combining activities with tabs
- Communicating with toast

Developing menus, animations, and options provide a way to navigate a user into the heart of your app, where the real work is accomplished. While menus depend upon button taps and finger swipes, they do not request and respond to input the same way that the core app functions do. Angry Birds, the popular Android game, begins with a menu to help the user select a level, but the real work of the app comes from reacting to the user's finger sliding the slingshot back and forth and then projecting one angry bird into some series of obstacles.

Up to now, you have explored ways to build the app skeleton, style it, and even animate it. In this chapter, you focus on adding the core functionality of the Simply Recipes app—adding new recipes by providing a form for user input. Creating the new recipe activity explores more advanced layout structures with ListViews and TabHosts, and you learn how to return some basic feedback with toast alerts.

Making a New Recipe

While the recipes app has come great lengths from Hello World, it still remains to perform the core function of any basic recipe application: make a new recipe. You can imagine some of the components required to build a successful recipe entry screen: inputs for name, description, ingredients, steps, and time to prepare among others.

This can be quite a lot to fit into a single Android view, so to make a more seamless user experience, the different components need to split across multiple layouts, which is combined together in code.

EditText and Other Widgets

In order to begin building a form that collects input, as opposed to simply reacting to inputs like clicking and tapping, you need more widgets than just buttons. Android provides support for a number of different widgets:

> **ANDROID DOES**
>
> In addition to all of the predefined widgets, layout, and view objects, Android allows you to create your own. Remember that these are just Java classes and can be recreated to your own sentiment.

- **EditText.** This widget class converts a TextView into an editable text field. This is the most basic way to enter text information. Unique XML attributes for EditText include:
  - **Hint.** Displays a recommendation for content. Frequently replaces a separate TextView label for the field.
  - **Lines.** The number of vertical rows to allow for content.
  - **InputType.** Defines an input method such as date, phone number, password, and auto-complete or correct.
  - **Text.** The text a user enters or has entered into the control.
- **CheckBox.** A check box that can either be one of two states: checked or unchecked. The XML **checked** attribute indicates the checked state of the control.
- **RadioButton.** Like a check box, a radio button can be one of two states: selected or not. RadioButton also uses the **checked** XML attribute. Unlike a check box, once selected a radio button cannot be unselected. Radio buttons are most frequently used in RadioGroups—collections of radio buttons.

- **ToggleButton.** Displays a checked state as a light indicator with checked state as "On" and unchecked as "Off." A toggle's primary attributes include:

  - **Checked.** Indicates whether the control is checked or unchecked.

  - **TextOn.** The text to display instead of "On."

  - **TextOff.** The text to display instead of "Off."

- **Spinner.** Displays a drop-down list of values, one of which can be selected. Spinner's unique XML attributes include:

  - **Entries.** An array of possible values for the list.

  - **Prompt.** The text to display at the top of the pick list.

Using a combination of these widgets, you can build a robust view to input information. As with other View subclasses, these widgets are typically first defined in XML. Within a ViewGroup, you can define inputs for a name and description (see Figure 7.1):

```
<EditText
android:id="@+id/name_new"
android:layout_width="fill_parent"
android:layout_height="wrap_content"
android:hint="Name"
android:lines="1"
/>
<EditText
android:id="@+id/description_new"
android:layout_width="fill_parent"
android:layout_height="wrap_content"
android:hint="Description"
android:lines="3"
/>
```

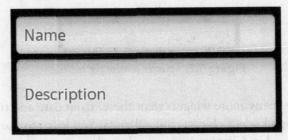

Figure 7.1: *Input for name and description.*

> **PITFALL**
>
> There is a lot of code for the remainder of the book, so from here on out we're focusing strictly on the layout—that is portrait/vertical—setup and not mentioning the layout-land—landscape/horizonal—mode. Feel free to add any layout code to the layout-land file, too.

Spinner controls (see Figure 7.2) are slightly more complicated and require an array resource for content and string resources for their prompt. Using the cuisine array from the options menu in Chapter 3, you can easily create a spinner with a selectable list of cuisines.

First ensure you have the array resource defined.

Then implement the control:

```
<TextView
    android:id="@+id/cuisine_display"
    android:layout_width="wrap_content"
    android:layout_height="wrap_content"
    android:text="Cuisine"
    android:paddingLeft="5dp"
/>
<Spinner
    android:id="@+id/cuisine_new"
    android:layout_width="fill_parent"
    android:layout_height="wrap_content"
    android:entries="@array/opt_cuisine_list"
    android:prompt="@string/mainnew_cuisine"
/>
```

As you can see from the Spinner control, the hint attribute is not available here, so a TextView positioned above provides the user with some information about the drop-down before it is selected. The Radio control is shown in Figure 7.3.

Figure 7.2: *Spinner control for cuisine.*

Android provides many more widgets than these, from date and time pickers to ratings bars, compound select objects, and galleries. You learn to use more widgets in the next chapter and in Chapter 10, but explore them on your own anytime you want to advance beyond the examples in the text.

Figure 7.3: *Radio control for cuisine.*

Deep Nested Views

As with any Activity whose layout becomes complex, the first step is to sketch a blueprint for the View. This screen should probably have a title, which is always visible at the top, with action buttons visible at the bottom. As the content of the recipe can include content from many widgets and be quite long, it needs to be scrollable. To achieve this, you can consider a layout like Figure 7.5.

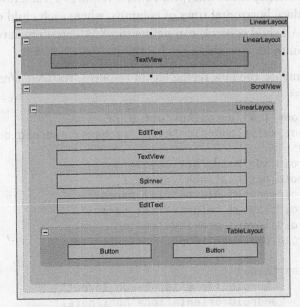

Figure 7.4: *Potential layout for the recipe View.*

While the TextView class handles its own scrolling, it is sometimes useful to nest ViewGroups between each other to maintain a certain aesthetic. In this case, the desire is to have a title always visible at the top of the screen and button widgets visible at the bottom.

A ScrollView enables this by designating a static location for scrollable content. ScrollViews can have only one child ViewGroup. Most frequently, this is a LayoutView, but can as easily be a TableView or RelativeView. These child layouts are free to have as many children as necessary.

With as much flexibility as Android offers, you can achieve the same effects by using different attributes of the available layouts. One of the standard ViewGroups, TableLayout, renders its child content in rows and columns. In this case, the use of a TableLayout is purely one of preference, but it does have some unique attributes worth exploring:

- **StretchColumns.** Specifies a column to stretch to available space in layout_width. "*" stretches all columns equally.

- **ShrinkColumns.** Specifies a column, "*" for all, to shrink.

- **CollapseColumns.** Specifies a column, "*" for all, to collapse.

In order for a TableLayout to interpret rows, it requires a TableRow subclass. TableRows produce horizontal rows of columns, where each child object inherits a new column. All child controls can define the XML attribute layout_column to identify the column for stretching or shrinking.

Using these elements, the new recipe layout resource, res/layout/recipe_tab_new.xml can be defined as:

```xml
<?xml version="1.0" encoding="utf-8"?>
<LinearLayout xmlns:android="http://schemas.android.com/apk/res/
    ➥android"
        android:id="@+id/root_layout"
        android:orientation="vertical"
        android:layout_width="fill_parent"
        android:layout_height="fill_parent"
        android:layout_gravity="center"
        android:padding="10dp"
        >
        <LinearLayout
            android:id="@+id/newsub_layout"
            android:orientation="horizontal"
```

```
                android:layout_width="fill_parent"
                android:layout_height="wrap_content"
                android:gravity="center"
                >
    <TextView
                android:id="@+id/new_search_button"
                android:layout_width="wrap_content"
                android:layout_height="wrap_content"
                android:text="Enter a New Recipe"
                android:textSize="20dp"
                />
    </LinearLayout>
    <ScrollView
                android:id="@+id/mainscroll_layout"
                android:layout_width="fill_parent"
                android:layout_height="wrap_content"
                android:paddingTop="20dp"
    >
    <LinearLayout
                android:id="@+id/recipe_table"
                android:layout_width="fill_parent"
                android:layout_height="wrap_content"
                android:orientation="vertical"
    >
    <EditText
                android:id="@+id/name_new"
                android:layout_width="fill_parent"
                android:layout_height="wrap_content"
                android:hint="Name"
                android:lines="1"
                />
    <EditText
                android:id="@+id/description_new"
                android:layout_width="fill_parent"
                android:layout_height="wrap_content"
                android:lines="3"
                android:hint="Description"
    />
    <TextView
                android:id="@+id/cuisine_display"
                android:layout_width="wrap_content"
                android:layout_height="wrap_content"
                android:text="Cusine"
                android:paddingLeft="5dp"
    />
```

continues

```
<Spinner
        android:id="@+id/cuisine_new"
        android:layout_width="fill_parent"
        android:layout_height="wrap_content"
        android:entries="@array/opt_cuisine_list"
        android:prompt="@string/main_menu_new_cuisine"
    />
<TableLayout
        android:id="@+id/recipe_table"
        android:layout_width="fill_parent"
        android:layout_height="wrap_content"
        android:orientation="horizontal"
        android:layout_gravity="center"
        android:gravity="center"
        android:stretchColumns="1"
        android:paddingTop="20dp"
    >
        <TableRow>
            <Button
            android:id="@+id/save_recipe"
            android:layout_width="fill_parent"
            android:layout_height="wrap_content"
            android:layout_column="1"
            android:text="Save"
            />
            <Button
            android:id="@+id/save_recipe"
            android:layout_width="wrap_content"
            android:layout_height="wrap_content"
            android:text="Cancel"
            />
        </TableRow>
    </TableLayout>
</LinearLayout>
</ScrollView>
</LinearLayout>
```

Implement the resource in a new Activity, RecipeEntry.java as:

```java
import android.os.Bundle;
public class RecipeEntry extends Activity {
    /** Called when the activity is first created. */
    @Override
```

```
public void onCreate(Bundle savedInstanceState) {
        super.onCreate(savedInstanceState);
        setContentView(R.layout.recipe_tab_new);
    }
}
```

The remaining steps to implement the new Activity should be familiar, beginning with Chapter 4, implementing the About Activity.

1. Add the Activity .RecipeEntry to AndroidManifext.xml.

2. Modify MainMenu onCreate to include a button click listener:

```
import android.widget.Button;
Button newRecipe = (Button)this.findViewById(R.id.main_new_ button);
newRecipe.setOnClickListener(this);
```

3. Modify MainMenu onCreateOptionsMenu() to add to the Options menu:

```
menu.findItem(R.id.main_menu_new).setIntent(
new Intent(this, RecipeEntry.class));
```

4. Modify MainMenu onClick() to associate the New button with the listener:

```
case R.id.main_new_button:
    Intent doMenuClick = new Intent(this, RecipeEntry.class);
    startActivity(doMenuClick);
    break;
```

Once complete, you can launch the app in the emulator to see the Enter a New Recipe screen in Figure 7.5.

GOOGLE IT

A popular philosophy for user interface design is Fitts' Law, which suggests that the closer and larger an item is, the easier it is to click. For this reason, sometimes seldom-used clickable items should be smaller and less easy to click. For example, users are more likely to want to save than cancel, so the Save button should be large and the Cancel button small.

Figure 7.5: *The Enter a New Recipe screen.*

Working with ListViews

What recipe app would be complete without lists? Recipes need ingredients, steps, measurements, and details which are not well suited to storage in large EditText controls. As with so many other obstacles, Android has anticipated this type of need and provided the *ListView* layout.

DEFINITION

The **ListView** class is treated as a widget and a ViewGroup. It extends the ListAdapter class, which it uses to bridge data back and forth.

Inflating the Layout

In order to display a known, static number of objects, any other type of ViewGroup would suffice. ListViews offer the potential to render dynamic content, which makes them quite powerful. A ListView can display a changing set of items, such as album covers or photos, which originate from another source. They can also expand to include new content directly.

To accomplish this feat, ListViews need a placeholder—to be nested in a View hierarchy just as any other layout type. They also need a layout definition for the content to inflate into them. The former can be created just as any other layout resource. See Figure 7.6.

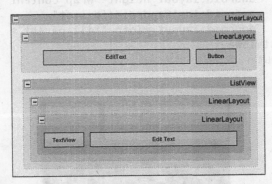

Figure 7.6: *The final layout hierarchy, once the layout and the layout inflation is built.*

Create the initial recipe_tab_ingredients.xml as (see Figure 7.7):

```xml
<?xml version="1.0" encoding="utf-8"?>
<LinearLayout
        android:id="@+id/ingredients_root"
        android:layout_width="fill_parent"
        android:layout_height="fill_parent"
        xmlns:android="http://schemas.android.com/apk/res/android"
        android:orientation="vertical"
        >
        <LinearLayout
                android:id="@+id/ingredients_child"
                android:layout_height="wrap_content"
                android:layout_width="fill_parent"
        >
        <EditText
                android:id="@+id/ingredient_text"
                android:layout_width="wrap_content"
                android:layout_height="wrap_content"
                android:layout_weight="1">
        </EditText>
        <Button
                android:id="@+id/ingredient_button"
                android:layout_width="wrap_content"
```

continues

```
            android:layout_height="wrap_content"
            android:text="Add Ingredient">
        </Button>
    </LinearLayout>
    <ListView
            android:layout_height="wrap_content"
            android:id="@android:id/list"
            android:layout_width="fill_parent">
    </ListView>
</LinearLayout>
```

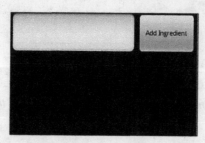

Figure 7.7: *The layout as rendered in Eclipse when you select the Graphical Layout tab. Notice that the child elements of the ListView are not yet visible.*

With the layout in place, it is time to build the inflation layout. This will be the pattern for all content nested in the ListView. Create recipe_tab_ingredients_inflate.xml as (see Figure 7.8):

```
<?xml version="1.0" encoding="utf-8"?>
<LinearLayout
        android:id="@+id/inflate_root"
        android:layout_width="fill_parent"
        xmlns:android="http://schemas.android.com/apk/res/android"
        android:orientation="vertical"
        android:layout_height="wrap_content"
        android:paddingLeft="5dp"
        >
        <LinearLayout
                android:id="@+id/inflate_sub"
                android:layout_height="wrap_content"
                android:layout_width="fill_parent"
                >
                <TextView
                        android:id="@+id/ingredient_cursor"
                        android:layout_width="wrap_content"
```

```
                    android:layout_height="wrap_content"
                    android:background="@drawable/arrow"
            >
            </TextView>
            <EditText
                    android:text="@+id/ingredient_first"
                    android:id="@+id/ingredient_first"
                    android:layout_height="wrap_content"
                    android:layout_width="fill_parent"
                    android:layout_weight="1"
            >
            </EditText>
        </LinearLayout>
</LinearLayout>
```

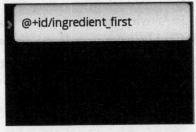

Figure 7.8: *The layout to inflate, as viewed in Eclipse. The text value for EditText will be informed through Java.*

 GOOGLE IT

If you want to use the drawable resources referenced in this text, feel free to use the source content for this code available at http://code.google.com/a/eclipselabs.org/p/cig-android-development, or Google images to find your own.

From XML to Java

With the proper resources in place, the ListView can be pieced together in a new Activity. This task is multifaceted, as you are not simply generating a static list of content. Rather, you are implementing a way to add ingredients to an ever expanding list.

Create the RecipeIngredients.java class, and begin implementing it with these imports:

```
import java.util.ArrayList;
import android.app.ListActivity;
import android.content.Context;
import android.os.Bundle;
import android.view.LayoutInflater;
import android.view.View;
import android.view.ViewGroup;
import android.view.View.OnClickListener;
import android.widget.BaseAdapter;
import android.widget.Button;
import android.widget.EditText;
import android.widget.ListView;
import android.widget.TextView;
```

The class should extend ListActivity instead of Activity as in your other activities. It should also implement OnClickListener in order to respond to button clicks.

```
public class RecipeIngredients extends ListActivity implements
  OnClickListener  {
For convenience, these class level properties will store values
  ➥needed between methods.

private static EditText _ingredientText = null;
private static Button _ingredientAdd = null;
private static final ArrayList<String> _ingredientListContents = new
  ➥ArrayList<String>();

@Override
public void onCreate( Bundle savedInstanceState ) {
      super.onCreate( savedInstanceState );
      setContentView( R.layout.recipe_tab_ingredients );
      ingredientListContents.add( "Salt" );
      ingredientListContents.add( "Pepper" );

      ingredientText=(EditText)findViewById( R.id.ingredient_text );
      ingredientAdd=(Button)findViewById( R.id.ingredient_button );
      ingredientAdd.setOnClickListener( this );
      setListAdapter( new ListViewAdapter(this) );
}
```

The communication of data to and from a ListView depends upon an adapter. ListViewAdapter() will call getView() to inflate the view with objects and data. It is in the class RecipeIngredients.

```
private static class ListViewAdapter extends BaseAdapter {
    private LayoutInflater ingredientInflater;

    public ListViewAdapter( Context context )
    {
    ingredientInflater = LayoutInflater.from( context );
    }
    public int getCount()
    {
    return _ingredientListContents.size();
    }
    public Object getItem(int position)
    {
    return position;
    }
    public long getItemId(int position)
    {
    return position;
    }
```

Once the required components are in place, implement getView() within the ListViewAdapter to take the inflation layout and prepare it for the ListView layout.

```
public View getView( int position, View view, ViewGroup group ) {

ListContent contents;

if (view == null)
{
        view = ingredientInflater.inflate( R.layout.recipe_tab_
➥ingredients_inflate, null );
        contents = new ListContent();
        contents.text = (EditText) view.findViewById( R.id.ingredient_
➥first );
        contents.text.setCompoundDrawables(view.getResources().
➥getDrawable(R.drawable.arrow_black), null, null, null );
        view.setTag(contents);
        }
else
{
        contents = (ListContent) view.getTag();
}

contents.text.setText( _ingredientListContents.get(position) );
```

continues

```
    return view;
    }

        static class ListContent {
            TextView text;
        }
    }
```

Finally, add the onClick() event to the RecipeIngredients class to capture new ingredient additions.

```
public void onClick(View v) {
        if( v == _ingredientAdd )
        {
        ingredientListContents.add( _ingredientText.getText().
    ➥toString() );
        setListAdapter( new ListViewAdapter(this) );
        }
        }
    }
```

If you now add the Activity to AndroidManifest.xml and modify the onClick() event in MainMenu, you can launch this activity in the emulator by clicking the **New Recipe** button:

```
case R.id.main_new_button:
        doMenuClick = new Intent(this, RecipeIngredients.class);
        startActivity(doMenuClick);
        break;
```

The ingredients list now renders the default ingredients, "Salt" and "Pepper", and you can add new ingredients to the list. With this ListView in place, you could modify this to include another EditText and Spinner picklist for measurements—tbsp, cup, or oz—to produce lines like "1 tbsp salt".

PITFALL

Adding fixed array values in code can be useful to demonstrate that the function works, but these should be removed as soon as testing has completed. You should avoid using hard-coded values in this way, unless they are truly necessary.

Simplify the Interface with TabHosts

One of the primary creative hurdles in mobile app development is working with the limited availability of screen real estate, that is, the amount of visible space on the mobile screen. Android tablets present the least obvious challenge, as 7 to 12 inches is usually more than adequate for more traditional interface designs. Android on Google TV boasts potentially huge screens, but users will sit many feet away from them—meaning your text needs to be large and visible.

User interfaces on mobile devices should be uncluttered, but you do not want users scrambling back and forth between screens to complete simple tasks. One of the ways to keep an interface simple and accessible is to use *tabs*.

DEFINITION

A **tab** is a visual marker that allows multiple resources to be contained in a single screen with a visual cue to distinguish them.

Using tabs, you can combine the recipe entry screen with the ingredients screen to produce a single, unified new recipe experience. As with ListViews, tabs are loosely defined in XML and primarily executed through Java. Tabs have three required classes:

- **Tabhost.** The parent ViewGroup. Its **id** must equal "tabhost".

- **TabWidget.** The first child Widget. Its **id** must equal "tabs".

- **FrameLayout.** The second child ViewGroup. Its **id** must equal "tabcontent".

Create recipe_tabs.xml as:

```xml
<?xml version="1.0" encoding="utf-8"?>
<TabHost xmlns:android="http://schemas.android.com/apk/res/android"
    android:id="@android:id/tabhost"
    android:layout_width="fill_parent"
    android:layout_height="fill_parent"
    >
    <LinearLayout
    android:id="@+id/tablinearlayout"
    android:orientation="vertical"
    android:layout_height="fill_parent"
    android:layout_width="fill_parent"
    >
```

continues

```
        <TabWidget
                android:id="@android:id/tabs"
                android:layout_height="wrap_content"
                android:layout_width="fill_parent">
        </TabWidget>
        <FrameLayout
                android:id="@android:id/tabcontent"
                android:layout_height="fill_parent"
                android:layout_width="fill_parent">
        </FrameLayout>
        </LinearLayout>
    </TabHost>
```

You may have noticed that the naming convention for the resources XML files in this chapter have all included "_tab". This provides an easy way to visually group the resource files in Package Explorer in Eclipse.

Tabs in Android function by bridging activities together in a new Activity class that defines the individual activities as new tabs. Thus, the RecipeEntry and RecipeIngredient classes are instanced as new tabs in a parent Activity class that extends TabActivity.

Individual tabs are not defined in XML; rather, they are instanced in the Activity itself. Tabs require the following elements:

- **TabHost.** A container for the tabbed window view. This can be instanced by defining:

```
TabHost _tabHost = getTabHost();
```

- **TabSpec.** A tab indicator with the content for the tab.

```
TabHost.TabSpec = _tabSpec;
```

Together, these objects can create tabbed content by calling the appropriate methods:

```
_tabSpec.newTabSpec()
.setContent();
    _tabHost.addTab( _tabSpec );
```

TabHost supports one important life cycle method, onTabChanged(). This method allows you to change the color and style of the currently selected and deselected tabs to provide a visual cue to the user on which tab is in focus.

With this information, you can build the unified new recipe screen as a new Activity class, RecipeNew as:

```
package com.recipesapp.basic;
import android.app.TabActivity;
import android.content.Intent;
import android.content.res.Resources;
import android.graphics.Color;
import android.os.Bundle;
import android.widget.TabHost;
import android.widget.TabHost.OnTabChangeListener;
public class RecipeNew extends TabActivity implements
   OnTabChangeListener {

TabHost _tabHost;
Resources _res;

@Override
public void onCreate( Bundle savedInstanceState ) {
      super.onCreate( savedInstanceState );
      setContentView( R.layout.recipe_tabs );

      tabHost = getTabHost();
      res = getResources();
      tabHost.setOnTabChangedListener( this );

      TabHost.TabSpec _tabSpec;

      tabSpec = _tabHost.newTabSpec("recipes").setIndicator(
   "Recipe", res.getDrawable( R.drawable.recipes_tab ) ).setContent(
   new Intent( this,RecipeEntry.class ) );
      tabHost.addTab( _tabSpec );

      tabSpec = _tabHost.newTabSpec("ingredients").setIndicator(
   "Ingredients", res.getDrawable( R.drawable.ingredients_tab ) )
.setContent( new Intent( this,RecipeIngredients.class ) );
      tabHost.addTab( _tabSpec );


      for(int i=0; i<_tabHost.getTabWidget().getChildCount(); i++)
      {
            tabHost.getTabWidget().getChildAt(i).
   setBackgroundColor( Color.LTGRAY );
      }

      tabHost.getTabWidget().setCurrentTab(1);
```

continues

```
        tabHost.getTabWidget().getChildAt(1).setBackgroundColor(
➡Color.DKGRAY );

        }


public void onTabChanged(String tabI(d){
for(int i=0; i<_tabHost.getTabWidget().getChildCount(); i++)
{
        tabHost.getTabWidget().getChildAt(i).setBackgroundColor(
Color.LTGRAY );
}

        tabHost.getTabWidget().getChildAt( _tabHost.getCurrentTab()
➡).setBackgroundColor( Color.GRAY );
}


}
```

Remember to add .RecipeNew to AndroidManifest.xml (see Figure 7.9) and change the onClick() event in MainMenu to call the RecipeNew class for R.id.main_new_button (see Figure 7.10).

Figure 7.9: *Adding a new recipe.*

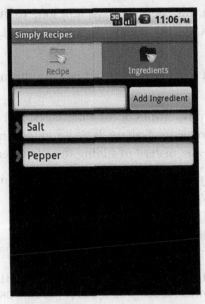

Figure 7.10: *Adding a new ingredient.*

Now the three activities are neatly linked together into a single screen, which appears to the user as a unified interface.

Toast Feedback

If you have ever filled out a long form on a website, you have probably entered a date or phone number in a format the site did not like. After reviewing a pop-up error message, you correct the formatting issue and move on. Android supports a similar kind of notification called a *toast*. For the Android, toast is a Widget class.

DEFINITION

A **toast** is a View containing a message to the user.

Toasts are very simple widgets, which require only a View to display and a length of time to run. A toast can take a View from an XML resource, or you can build a simple View in Java. A toast has only a few methods:

- **setView().** This gives the toast some view to display.

- **setDuration().** Instructs the toast how long to display, usually either Toast. LENGTH_LONG for a long interval or Toast.LENGTH_SHORT for a short time.

- **cancel().** Used to end the toast if it is still running.

To start exploring the possibilities of toast messages, you can add some validation logic to the RecipeIngredients Activity. If a user enters an invalid character, such as "/", display a toast message instead of adding the ingredient to the list.

Start by creating a simple method in the RecipeIngredients class:

```
import android.graphics.Color;
import android.view.Gravity;
import android.widget.Toast;
public void invalidCharacterToast(String character)
{
        TextView displayView = new TextView(this);
        displayView.setBackgroundColor(Color.DKGRAY);
        displayView.setTextColor(Color.RED);
        displayView.setPadding(10,10,10,10);
        displayView.setText(character + " is an invalid character");
        Toast theToast = new Toast(this);
        theToast.setView(displayView);
        theToast.setDuration(Toast.LENGTH_LONG);
        theToast.setGravity(Gravity.CENTER_VERTICAL, 0,0);
        theToast.show();
}
```

This creates a basic TextView and displays it as a toast. Now, to add some logic to the onClick() event (see Figure 7.11):

```
public void onClick(View v) {

String IngredientText;
if( v == _ingredientAdd )
{
IngredientText = _ingredientText.getText().toString();
        if(IngredientText.contains( "/" ))
                invalidCharacterToast("/");
        else
                ingredientListContents.add( _ingredientText.getText().
    toString() );
        setListAdapter( new ListViewAdapter(this) );
        }
}
```

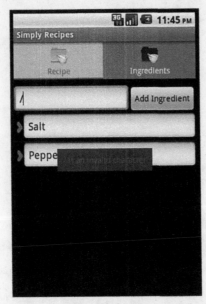

Figure 7.11: *Now, if the user enters a "/" character, a toast will prompt them to change their ways.*

With widgets and views and toasts, you can build robust user interfaces. You have put a lot of XML and code together to create a dynamic user interface, which you extend in the next chapter to actually store and edit data. In the next chapter, you explore how to tune this interface to use for both entry of new records as well as management of existing recipes.

The Least You Need to Know

- ScrollViews allow long content to scroll on screen.
- Unlike other ViewGroups, ListViews and TabHosts require both XML and Java to fully function.
- Spinners deliver drop-down lists of content.
- Use toasts to communicate messages to the user.

Figure 7.11   ... the user enters a ... ViewList.4 reset, and groups grow as ... change their color.

With widgets and views and toasts, you can build robust user interfaces. You have put a lot of XML and code together to create a dynamic user interface, which you started in the last chapter to actually store and edit data. In the next chapter, you explore how to tune this interface to use for both entry of new records as well as management of existing records.

The Least You Need to Know

- ScrollViews allow long content to scroll on screen.
- Unlike other ViewGroups, ListViews and TableHosts require both XML and Java to fully function.
- Spinners deliver drop-down lists of content.
- Use toasts to communicate messages to the user.

Storing and Retrieving Data

In This Chapter

- Introduction to SharedPreferences
- Possibilities of local storage
- Working with external data
- Saving and reading recipes in your app

Of all the apps you have opened more than once on any device, how many remember nothing about you? Very few. Reopening the solitaire app takes you back to the last game you started. Opening the phone app returns you to your last call. More than that, these apps remember not just the last activity you used, but frequently all your activities. Photo apps would not be very useful if they did not collect and store all the pictures you took with the camera.

When you open the Contacts app, you expect it to contain the same information as the last time you used it. From settings to content, it should all be there, ready for use. Up to this point, you have invested a lot of energy in developing the user interface and improving the user experience. In this chapter, you begin storing the information that users enter. With stored data, users can manage recipes, share them, search for them, and do everything else you can imagine to develop.

Storage Overview

The days of rotary phones are nearly ancient history from a technology perspective. To dial a number, you opened a phone book, found a contact, and then painfully rotated the phone's dial until a number would ring. Digital phones added quick dial, a primitive form of address book; but it was not until mobile phones arrived that contacts and dialing were truly unified.

Storage is one of the key components that made this obvious technological leap possible. Few users now want rotary-styled apps. Instead, apps should provide the shortest possible path to completing an activity by storing information that users enter. Android simplifies this process by providing access to multiple, complimentary kinds of storage.

SharedPreferences

In Chapter 4, you created an Options menu that collects some preferences about the Simply Recipes app. These preferences are managed by the *SharedPreferences* class, which provides Activities a mechanism to store key/value information such as preferences or state. Keys identify the name of the preference, while value represents its string, integer, or Boolean value.

> **DEFINITION**
>
> **SharedPreferences** are stored as an XML resource. By default, this resource is private and can only be accessed by the calling activity.

Preferences take three forms:

- **Shared.** These preferences can be shared across all components in an app or across all apps. These are accessed by calling:

```
getSharedPreferences( Preference, Mode );
```

"Preference" is a String value that must exactly match the name of the preference.

"Mode" is an enum on the Activity:

- **MODE_PRIVATE.** Accessible only to the current app.

- **MODE_WORLD_READABLE.** Visible to all apps on the device.

- **MODE_WORLD_WRITABLE.** All apps can modify this value.

- **Private.** These preferences are visible only to the Activity that sets them. These are accessed by calling:

```
getPreferences( Mode );
```

- **Private for Inactivity.** These preferences are visible only to the current Activity, and only when the Activity is inactive. These preferences are lost by clicking **Back** or closing the Activity.

In the onSaveInstanceState() method, you can save the current View information by calling:

```
outState.putString( Preference, CurrentView.toString() );
```

In the onRestoreInstanceState() method, you can restore the previous View information by calling:

```
CurrentView.setText( inState.getString( Preference ) );
```

In general, you use getPreferences() and getSharedPreferences() to interact with preference information. To interact with either of these methods, you need to use the SharedPreferences Editor. The Editor uses a few methods to modify preferences:

- **putString().** Places a string preference into the collection.

- **putInt().** Places an integer into the collection.

- **remove().** Removes a preference from the collection.

- **commit().** Saves the modifications. No changes will be saved without a commit().

To put this all together and set and retrieve a preference, you can instrument a class with these starter methods:

```
private static final String AboutPreference =
    "AboutLanguagePreference";
private void savePreferences() {
    SharedPreferences aboutSharedPreferences = getSharedPreferences
    (AboutPreference, About.MODE_PRIVATE);
    SharedPreferences.Editor aboutEdit = aboutSharedPreferences.
    edit();
    aboutEdit.putString( "Language", "English" );
    aboutEdit.commit();
}

private void restorePreferences() {
    SharedPreferences aboutSharedPreferences = getSharedPreferences
    (AboutPreference, About.MODE_PRIVATE);
```

continues

```
        String language = aboutSharedPreferences.getString( "Language",
→null );
    }
```

SharedPrefences provides a powerful way to keep consistent information about an app, from what the user last did to which localized settings should apply.

ANDROID DOES

SharedPreferences are for more than just storing traditional preferences, like the user's favorite color. There is some overlap, as the PreferenceActivity does store user preferences using SharedPreferences.

While SharedPreferences offers an excellent way to deal with key/value pairs, it is not an efficient way to manage a large volume of data. For this purpose, another type of storage is required.

Internal Storage

The Android OS is a Linux OS, which means that it has a file system that both Android and its apps can access. Almost all Android devices use reliable flash memory, which means that once a file is written, the file is safe from harm. Internally, you can open your /raw resource files for read access; but for security purposes, each app has write access to a single, private directory /Android/data/<your_package_name>/ files.

After importing the appropriate Java packages, namely java.io.FileOutputStream and java.io.FileInputStream, you can access the methods needed to access the file system. These methods are described in the following table.

File Operation Methods

Access	Method	Description
Write	**openFileOutput**(Filename, Mode)	Creates or opens a file with the name of Filename and opens it according to the Mode—MODE_PRIVATE is the default.

Access	Method	Description
Write	**write()**	Commits changes to the file to disk.
Read	**openFileInput(Filename)**	Opens file for read access.
Read	**read()**	Reads from the open file.
Read	**openRawResource()**	Opens a file from the /res/raw directory.
Read/Write	**close()**	Ends communication with the file.
Info	**fileList()**	Returns a string array of all files in the local directory.
Info	**getFilesDir()**	Returns the path to your local files.
Delete	**deleteFile()**	Permanently deletes a file.

Additionally, you can fetch the Android cache directory by calling getCacheDir(). You can write nonpersistent, temporary files here if you need to save noncritical information for quick access later.

PITFALL

Files stored in cache are deleted if Android needs more temporary space or your app is uninstalled. Be sure to use cache space sparingly and save important data elsewhere.

Putting it all together, you can create a new file or open an existing file of the same name and give it some data by implementing this demonstration method:

```
private void saveItToFile(String filename, String content){
    try
        {
        FileOutputStream fos = openFileOutput(filename, About.
   ➡MODE_PRIVATE);
        fos.write(content.getBytes());
        fos.close();
        }
        catch(IOException e){}
    }
```

With file system access, you can read and write any valid data to and from disk. Unfortunately, internal storage space is limited. Many Android devices ship with just 8 gigabytes of free space. Your user may have filled this with songs and photos before your app ever got a chance to write a file.

External Data

External storage is a standard Android feature. Most devices include SD card slots that expand the available storage space by many gigabytes. Accessing this space comes with some conditions.

- The user can remove the SD card or extra storage at any time. Android provides no security mechanism to protect data that is not yet written to disk.

- Files stored on external storage are directly visible to all other apps and the user.

- If a user deletes or uninstalls your app, the only files that Android preserves are on external storage.

The same methods apply to working with files on external resources, but you first need to know what directories to use. As of Android 2.2, the getExternalFilesDir() method will instruct you. This method takes a type, such as:

- **DIRECTORY_MUSIC.** The standard folder for music.

- **DIRECTORY_DOWNLOADS.** The default download folder.

- **DIRECTORY_PICTURES.** Directory for photos.

- **null.** If no type is supplied, the root directory is returned.

ANDROID DOES

To keep Media Scanner from indexing your external files, create an empty file named ".nomedia" in your external files directory. This will keep your files out of apps like Music or Gallery.

Saving a New Recipe

Now that you have a few tools to take user input and save it, it is time to implement some new features in the Simply Recipe app. First, the Enter a New Recipe screen needs to save the recipe when the Save button is clicked. Second, you need to implement a List Recipes screen to navigate stored recipes.

Before beginning to design the interfaces, you need to decide how to store this data. Ironically, Android has not implemented a reliable way to work with XML files generated by individual apps. This is unfortunate, as XML is an ideal way to work with data sets. Given this limitation, it quickly becomes a parsing nightmare to read and write data from individual files on the filesystem.

You learn how to use Structured Query Language (SQL) to solve this problem in Chapter 17, but for now SharedPreferences meets the need. SharedPreferences, while limited to key/value pairs, are stored in XML and will meet the immediate need to save and retrieve recipes.

Retrieving SharedPreferences

As you confront tasks in Android app development that are increasingly complex, it becomes useful to write a short *specification* with the steps involved to complete the project.

DEFINITION

A **specification** (spec) is a set of requirements for accomplishing a given task. Specs are as detailed and explicit as required for the person performing the task to understand and for peers to understand as well.

Here is the specification for our recipe app:

1. Create recipe_list.xml and recipe_list_inflate.xml resources to define a ListView and inflation content.

2. Create a new Activity class **RecipeList** which extends ListActivity and implements OnItemClickListener.

3. **RecipeList** inflates a ListView with all recipes the user has entered.

4. Clicking a recipe in **RecipeList** launches the **RecipeNew** Activity to view and edit the recipe.

5. Add a button to **MainMenu** to launch the **RecipeList** Activity.

You should refine these steps to meet your needs—more or less specific details may be required to make this spec right for you. See Figure 8.1.

Figure 8.1: *The way our app will look with searchable recipes.*

Step one should be familiar from Chapter 7. The ListView resources needed here will be very similar to those in the RecipeIngredients Activity. Define recipe_list.xml as:

```
<?xml version="1.0" encoding="utf-8"?>
<LinearLayout xmlns:android="http://schemas.android.com/apk/res/
    android"
        android:id="@+id/list_root"
        android:layout_width="fill_parent"
        android:layout_height="fill_parent"
        android:orientation="vertical"
    >
        <LinearLayout
            android:id="@+id/list_child"
            android:layout_height="wrap_content"
            android:layout_width="fill_parent"
            android:layout_gravity="center"
            android:gravity="center"
            android:paddingTop="10dp"
            android:paddingBottom="10dp"
        >
```

```
<TextView
        android:id="@+id/list_header"
        android:layout_width="wrap_content"
        android:layout_height="wrap_content"
        android:text="Your Recipes"
        android:textSize="20dp"
>
</TextView>
</LinearLayout>
<ListView
        android:id="@android:id/list"
        android:layout_height="wrap_content"
        android:layout_width="fill_parent">
</ListView>
</LinearLayout>
```

This simple layout defines a title for the screen and a ListView placeholder. Next, create recipe_list_inflate.xml as:

```
<?xml version="1.0" encoding="utf-8"?>
<LinearLayout xmlns:android="http://schemas.android.com/apk/res/
    android"
        android:id="@+id/inflate_root"
        android:layout_width="fill_parent"
        android:orientation="vertical"
        android:layout_height="wrap_content"
        android:paddingLeft="5dp"
>
    <LinearLayout
    android:id="@+id/inflate_sub"
    android:layout_height="wrap_content"
    android:layout_width="fill_parent"
    android:orientation="horizontal"
    android:paddingBottom="5dp"
    >
        <TextView
                android:id="@+id/list_cursor"
                android:layout_width="wrap_content"
                android:layout_height="wrap_content"
                android:background="@drawable/arrow"
                android:paddingRight="10dp"
        >
        </TextView>
        <TextView
                android:text="@+id/recipe_first"
```

continues

```
                              android:id="@+id/recipe_first"
                              android:layout_height="wrap_content"
                              android:layout_width="fill_parent"
                              android:textSize="20dp"
                              android:layout_weight="1"
                       >
                       </TextView>
               </LinearLayout>
       </LinearLayout>
```

Next, you create the RecipeList activity to drive the list of recipes. Because SharedPreferences are stored as key/value pairs, some care is needed when deciding to use them for storage of dynamic data. A shared preference must be called by name, which means that the names you choose for preferences should change only in predictable, programmatic ways.

To ensure that the core preference names do not inadvertently change, declare them as public static final String values, which will effectively make them constant. In MainMenu, declare these variables:

```
public static final String RecipeNamesPref = "RecipeNames";
public static final String NamePref = "Name";
public static final String CuisinePref = "Cuisine";
public static final String DescriptionPref = "Description";
public static final String StepsPref = "Steps";
public static final String SelectedRecipe = "SelectedRecipe";
```

In this way, you can be assured that all references to MainMenu.RecipeNamesPref always return the same, constant value "RecipeNames". If you visualize each shared preference as a unique record, which can have as many child key/values pairs as are needed, you can imagine each recipe as its own shared preference.

As a shared preference, each recipe needs to be called by name. To make it easy to remember all of the recipe names a user has made, you will create a shared preference for the whole collection of recipe names.

Begin by implementing the RecipeList Activity class as:

```
package com.recipesapp.basic;
import java.util.ArrayList;
import android.app.ListActivity;
import android.content.Context;
import android.content.Intent;
import android.content.SharedPreferences;
import android.content.SharedPreferences.Editor;
```

```
import android.os.Bundle;
import android.view.LayoutInflater;
import android.view.View;
import android.view.ViewGroup;
import android.widget.AdapterView;
import android.widget.BaseAdapter;
import android.widget.ListView;
import android.widget.TextView;
import android.widget.AdapterView.OnItemClickListener;
public class RecipeList extends ListActivity implements
➡OnItemClickListener  {
```

Much of the following code should be familiar from Chapter 7, but several key components have changed. Follow the code comments to interpret the updates.

```
    //Store the list of recipes as an ArrayList
private static final ArrayList<String> _RecipeListContents = new
➡ArrayList<String>();
private ListView recipeListView;

@Override
    public void onCreate( Bundle savedInstanceState ) {
            super.onCreate( savedInstanceState );
            setContentView( R.layout.recipe_list );

            //Retrieve the string containing all recipe names from
➡SharedPreferences
            SharedPreferences recipeNames = getSharedPreferences(
➡MainMenu.RecipeNamesPref, RecipeEntry.MODE_WORLD_READABLE);

            //Convert the string list of recipes into an array
            String[] recipeList = recipeNames.getString( MainMenu.
➡RecipeNamesPref, "Make New" ).split( "," );

            //For each recipe in the list, add it to the ArrayList
➡of recipes
            for(String recipe: recipeList)
            {
                    RecipeListContents.add( recipe );
            }

            //Instance the recipe list ListView
            recipeListView = (ListView) findViewById(android.R.id.
➡list);
            recipeListView.setAdapter( new ListViewAdapter(this) );
```

continues

```
                    recipeListView.setTextFilterEnabled(true);

                    //Add an OnItemClickListener to respond to recipe
            selections
                    recipeListView.setOnItemClickListener( this );

                    }

                    //setListAdapter( new ListViewAdapter(this) );
                    private static class ListViewAdapter extends
            BaseAdapter {
                    private LayoutInflater recipeInflater;

                    public ListViewAdapter( Context context )
                    {
                            recipeInflater = LayoutInflater.from( context );
                    }
                    //Required method
                    public int getCount()
                    {
                            return _RecipeListContents.size();
                    }
                    //Required method
                    public Object getItem(int position)
                    {
                            return position;
                    }
                    //Required method
                        public long getItemId(int position)
                    {
                            return position;
                    }
                    public View getView( int position, View view, ViewGroup
            group ) {

                    ListContent contents;

                    if (view == null)
                            {
                            view = recipeInflater.inflate( R.layout.recipe_
            list_inflate, null );
                            contents = new ListContent();
                            contents.text = (TextView) view.findViewById(
            R.id.recipe_first );
                            contents.text.setCompoundDrawables(view.
```

```
            getResources().getDrawable(R.drawable.arrow_black), null, null,
        null );
                    view.setTag(contents);
            }
            else
            {
                    contents = (ListContent) view.getTag();
            }

                    contents.text.setText( _RecipeListContents.
        get(position) );
                    return view;
            }

            static class ListContent {
                    TextView text;
            }
    }
    //When a list item is clicked
    public void onItemClick( AdapterView<?> arg0, View arg1, int
        arg2, long arg3 ) {

            //Set the selected recipe for the next Activity to use
            SharedPreferences selectedRecipe =
        getSharedPreferences( MainMenu.SelectedRecipe, RecipeEntry.MODE_
        WORLD_READABLE);
            Editor selectedEdit = selectedRecipe.edit();
            selectedEdit.putString( MainMenu.SelectedRecipe, _
        RecipeListContents.get( arg2 ) );
            selectedEdit.commit();

            //Start the RecipeNew activity
            Intent doRecipeClick = new Intent(this, RecipeNew.
        class);
            startActivity(doRecipeClick);
    }

}
```

SharedPreferences are stored as single entity values—a single string or a single integer. Frequently, you want to use collections, in which case you must convert the raw value into an appropriate collection or *array*.

> **DEFINITION**
>
> **Arrays** are systematic collections of objects, usually arranged serially in rows and columns.

Java supports a number of types of collections. Traditional arrays are instanced by defining:

```
String[] newStringArray = {"Cake","Pudding"};
```

Which creates a single column array with two rows:

```
newStringArray[0] == "Cake"
newStringArray[1] == "Pudding"
```

Arrays provide an easy way to serialize and deserialize data; they convert simple values into rows and columns that can be organized. Java also supports a powerful companion to traditional arrays, the ArrayList. These can be defined by naming:

```
ArrayList<String> newArrayList = new ArrayList<String>();
```

These lists are optimized array instances for a type, such as String or Integer; and they support additional methods:

- **add().** Places a reference into the ArrayList.

- **get().** Retrieves a reference from the ArrayList by index position.

- **size().** Returns the size of the ArrayList.

- **remove().** Removes the reference at the index position and shifts subsequent elements to the left.

- **indexOf().** Returns the index position of the first occurrence of the reference.

- **clear().** Removes all references from the ArrayList.

Storing Input as SharedPreferences

The RecipeList Activity is now prepared to list the recipes a user has entered. Now, RecipeEntry must be modified to collect and store new recipes and changes to recipes. It should also be modified to distinguish between new entries and edits. Begin by creating a saveRecipe() method in RecipeEntry. This will do the heavy lifting of storing the data and can be called by the onClick() event:

```
public void saveRecipe() {

        //Create or Open a Preference for all Recipe Names
        SharedPreferences recipeName = getSharedPreferences( MainMenu.
RecipeNamesPref, RecipeEntry.MODE_WORLD_READABLE);
        //Get the recipe list. If empty, the list will contain only
"New Recipe"
        String recipeList = recipeName.getString( MainMenu.
RecipeNamesPref, "New Recipe" );
        SharedPreferences.Editor recipeNameEdit = recipeName.edit();
        EditText newRecipeNameView = (EditText)this.findViewById( R.id.
name_new );
        String newRecipeName = (String)newRecipeNameView.getText().
toString();

        //We must have a name in order to save
        if( null != newRecipeName )
        {
                //Preserve the existing list and append the new name
                recipeList = recipeList + "," + newRecipeName;
                recipeNameEdit.putString( MainMenu.RecipeNamesPref,
recipeList );
                recipeNameEdit.commit();

                //Create or open the new recipe as a SharedPreference:
<name>_detail to identify the recipe record
                String newRecipeDetail = newRecipeName + "_Detail";
                SharedPreferences newRecipe = getSharedPreferences(
newRecipeDetail, RecipeEntry.MODE_WORLD_READABLE);
                SharedPreferences.Editor newRecipeEdit = newRecipe.
edit();
                //Store the recipe data for this recipe
                newRecipeEdit.putString( MainMenu.NamePref,
newRecipeName );
                newRecipeEdit.putString( MainMenu.DescriptionPref,
((EditText)this.findViewById( R.id.description_new )).getText().
toString() );
                newRecipeEdit.putInt( MainMenu.CuisinePref, ((Spinner)
this.findViewById( R.id.cuisine_new )).getSelectedItemPosition() );
                newRecipeEdit.putString( MainMenu.StepsPref,
((EditText)this.findViewById( R.id.steps_new )).getText().toString() );
```

continues

```
                    newRecipeEdit.commit();

                }

        }
```

With the method in place, it just needs to be called in response to a user clicking the
Save button. Take the opportunity to give some action to the Cancel button as well.
Once clicked, Save should store the recipe in SharedPreferences and return the user
to the Main Menu. Cancel should just exit the screen to Main Menu. In onCreate(),
add onClickListeners to the buttons:

```
        Button saveButton = (Button)this.findViewById( R.id.save_new_
➥recipe );
        saveButton.setOnClickListener( this );
        Button cancelButton = (Button)this.findViewById( R.id.cancel_
➥new_recipe );
        cancelButton.setOnClickListener( this );
```

Implement the onClick() method:

```
        public void onClick(View thisView) {
                Intent doMenuClick;
                switch (thisView.getId()) {
                case R.id.save_new_recipe:
                //Save the recipe first!
        saveRecipe();
                doMenuClick = new Intent( this, MainMenu.class );
                startActivity( doMenuClick );
                break;
        case R.id.cancel_new_recipe:
                doMenuClick = new Intent( this, MainMenu.class );
                startActivity( doMenuClick );
                break;
        }
        }
```

At this point, the app compiles and runs; but, if you click a recipe from the new
RecipeList Activity you see only an empty, "Enter a New Recipe" screen. Fear not!
The data has not been lost, but the recipe information must be mapped into the view.
Add one more statement to MainMenu's onCreate():

```
    SharedPreferences selectedRecipe = getSharedPreferences(
➡MainMenu.SelectedRecipe, RecipeEntry.MODE_WORLD_READABLE);
    SelectedRecipe = selectedRecipe.getString( MainMenu.
➡SelectedRecipe, "New Recipe" );

    if("New Recipe" != _SelectedRecipe)
    {
        SharedPreferences thisRecipe = getSharedPreferences(
➡_SelectedRecipe + "_Detail", RecipeEntry.MODE_WORLD_READABLE);
        EditText NameText = (EditText)this.findViewById( R.id.
➡name_new );
        NameText.setText( thisRecipe.getString( MainMenu.
➡NamePref, "" ) );
        EditText DescriptionText = (EditText)this.findViewById(
➡R.id.description_new );
        DescriptionText.setText( thisRecipe.getString(
➡MainMenu.DescriptionPref, "" ) );
        EditText StepsText = (EditText)this.findViewById( R.id.
➡steps_new );
        StepsText.setText( thisRecipe.getString( MainMenu.
➡StepsPref, "" ) );
        Spinner CuisineSelect = (Spinner)this.findViewById(
➡R.id.cuisine_new );
        CuisineSelect.setSelection( thisRecipe.getInt(
➡MainMenu.CuisinePref, 0 ));
        TextView title = (TextView)this.findViewById( R.id.edit_
➡recipe_title );
        title.setText( "Edit This Recipe" );
    }
```

Finally, all of the components are in place. When a user adds a new recipe, it is added to the list of stored recipes. That list is now visible from the RecipeList screen, which directs the user back to edit the recipe once selected.

Manage Your Recipes

If it were not for the spec, you might have saved, compiled, and run the app only to scratch your head repeatedly, wondering "Where is the recipe list screen?" How do you see the recipes? Luckily, the printed spec sits next to your keyboard to remind you: add the Recipe List button to the Main Menu.

Sometimes the simplest steps are the easiest to forget. Adding the button and onClick() event should now be old hat. Modify main.xml to read:

```xml
<?xml version="1.0" encoding="utf-8"?>
<LinearLayout xmlns:android="http://schemas.android.com/apk/res/
  android"
      android:id="@+id/root_layout"
      android:orientation="vertical"
      android:layout_width="fill_parent"
      android:layout_height="fill_parent"
      android:layout_gravity="center"
      android:padding="10dp"
      >
      <LinearLayout
              android:id="@+id/mainsub_layout"
              android:orientation="horizontal"
              android:layout_width="fill_parent"
              android:layout_height="wrap_content"
          >
              <Button
                      android:id="@+id/main_search_button"
                      android:layout_width="wrap_content"
                      android:layout_height="wrap_content"
                      android:text="@string/main_search_button"
                  />
              <EditText
                      android:id="@+id/main_search"
                      android:layout_width="fill_parent"
                      android:layout_height="wrap_content"
                      android:lines="1"
                  />
      </LinearLayout>
      <ScrollView
              android:layout_width="fill_parent"
              android:layout_height="wrap_content"
              android:paddingTop="20dp"
          >
      <LinearLayout
              android:id="@+id/mainscroll_layout"
              android:orientation="vertical"
              android:layout_width="fill_parent"
              android:layout_height="wrap_content"
          />
```

```
                    <Button
                            android:id="@+id/main_new_button"
                            android:layout_width="fill_parent"
                            android:layout_height="wrap_content"
                            android:text="@string/main_new_button"
                    />
                    <Button
                            android:id="@+id/main_list_button"
                            android:layout_width="fill_parent"
                            android:layout_height="wrap_content"
                            android:text="List Recipes"
                    />
                    <Button
                            android:id="@+id/main_help_button"
                            android:layout_width="fill_parent"
                            android:layout_height="wrap_content"
                            android:text="@string/main_help_button"
                    />
                    <Button
                            android:id="@+id/main_about_button"
                            android:layout_width="fill_parent"
                            android:layout_height="wrap_content"
                            android:text="@string/main_about_button"
                    />
                    </LinearLayout>
                </ScrollView>
        </LinearLayout>
```

Then set the Button's onClickListener in MainMenu onCreate():

```
Button listRecipe = (Button)this.findViewById(R.id.main_list_button);
listRecipe.setOnClickListener(this);
```

After adding the onClick() event:

```
            case R.id.main_list_button:
                    doMenuClick = new Intent(this, RecipeList.class);
                    startActivity(doMenuClick);
                    break;
```

You can run the app and begin working with stored data. See Figure 8.2 and Figure 8.3.

Figure 8.2: *The list of recipes.*

Figure 8.3: *Editing a recipe description.*

Allow a deep sense of satisfaction to settle as you have taken the Simply Recipes app one step closer to completion. You have probably noticed aesthetic choices which need to be changed to improve the user interface, and you have probably anticipated shortcomings with the sample code which need to be improved to accommodate different behaviors: How to delete a recipe? Prevent duplicates? Edit many recipes at the same time?

Be encouraged that you already have the tools necessary to solve these problems and improve the app experience for your users! In the next chapter, you learn how to start implementing search against the data in your app.

The Least You Need to Know

- SharedPreferences provide XML storage for key/value pairs locally for your Activity or globally.
- Local storage provides read and write access to a private directory for your app.
- Include /res/raw resources for read-only file access.
- Use external storage, such as on SD cards, to store public files and content.
- All files in local storage are removed if your app is uninstalled.

Search for It

In This Chapter

- Introduction to the Quick Search Box
- Making your app searchable
- Process the search results
- Implement voice search

When you first think of Google, you probably think "search." The two words are used almost interchangeably, so much so that users of other search engines such as Bing and Yahoo! will refer to "searching" as "googling." It should then come as no surprise that search is integral to the Android experience. In addition to the required physical buttons for Home, Menu, and Back, Search is standard on most devices.

But how does search really affect the Android experience? This chapter explores how search in Android works, what content is included, and how to extend search capabilities into your own app.

Search Basics

Android provided an integrated search experience via the Quick Search Box, a search dialog called when users request a search action. With Search, users can find content on their device or from the Internet. A search for "Doe" could return Android contacts with the name, a musical app, and Internet results relating to deer.

Increasingly, users expect to be able to search all of the content on their devices. Adding search capability is an important step in the process of developing a well-rounded app.

A search request opens an instance of the search dialog. In order to maintain a consistent look and feel across each device, Android manages the entire life cycle of the search dialog with the Search Manager class. You need only create the appropriate resources, and Android will create the dialog, execute the search, and return any selection, if applicable, back to your app. See Figure 9.1.

Figure 9.1: *A search for "apple pie."*

Once you have enabled search within your app, you will be able to provide some of these features to your users:

- Allow search of all your app's content
- Enable voice search
- Use recent searches to provide search recommendations
- Use data from your app to offer search suggestions
- Provide your apps search suggestions to the global search dialog

As always, you will want to draft a spec for implementing these new features that begins by answering:

- What content from my app will be available to search?
- How much of this will be available to search from other apps?
- How will my app allow users to control their privacy?

> **PITFALL**
>
> Whenever your app exposes information to other Android apps, you should always consider the impact on a user's privacy. Many users consider their search history as something that should be private by default. Always clearly communicate what information is shared outside your app and provide a way for users to opt-out.

After you have decided what to search, it is time to implement the Search Manager in your app. You will need to:

1. Add a searchable configuration.
2. Add a searchable Activity.
3. Add a control to the Activity to launch the search dialog.

Create the Configuration

Configuration resources inform Android which apps have incorporated search and where. For search, this file is usually named searchable.xml and must be placed in the /res/xml project folder. You are free to provide any name for the file, but it must include the <searchable> tag and define the android:label attribute.

```
<?xml version="1.0" encoding="utf-8"?>
<searchable
        xmlns:android="http://schemas.android.com/apk/res/android"
        android:label="Simply Recipes Search"
        android:hint="Search for Your Recipes and Ingredients" >
</searchable>
```

Additional attributes are available that provide configuration control to different search features. These include th following.

Common Searchable Attributes

Affects	Attribute	Description
Global Search	android:includeInGlobalSearch	Set to "true" if your app's search should be included in the Quick Search Box.
Global Search	android:searchSettingsDescription	Defines a brief description for the kind of results returned from your app.
Suggestions	android:searchSuggestAuthority	This attribute is necessary to provide search suggestions. This string value should match a ContentProvider defined in your AndroidManifest.xml.
Suggestions	android:searchSuggestIntentAction	Defines an Intent action to trigger when a user clicks on a custom search suggestion.
Suggestions	android:searchSuggestIntentData	Defines an Intent data to include with the default Intent action when a user clicks on a custom search suggestion.
Suggestions	android:searchSuggestThreshold	Defines the number of characters after which to begin offering suggestions. Default is 0, which means suggestions will start immediately.

Affects	Attribute	Description
Voice Search	android:voiceSearchMode	Enables voice search for the device.
Voice Search	android:voicePromptText	Defines an additional message to display in the input dialog.

As you configure search for your app, you will see frequent references to Content-Providers, which are discussed in full in Chapter 18. Some aspects of search are dependent on having implemented a ContentProvider. Feel free to process this information now by skipping ahead, or later by revisiting the relevant sections.

Define a Searchable Activity

When a user requests to search, the Search Manager launches the search dialog box. The Search Manager automatically handles search suggestions, but as soon as the user executes the search, the Search Manager needs an Activity to process and display the full result set of the search.

Your AndroidManifest.xml must define an Activity as searchable and provide this search context to other activities in the app. We'll call the search action "intent-filter". To do this, follow these steps:

1. Add the new activity to the manifest as the searchable Activity. This requires defining a search action as an <intent-filter> element. When a user requests a search within your app, Android needs to know which Activity to associate with the search results. You must also add the searchable configuration to the <meta-data> element.

```
<activity android:name=".RecipeSearch" >
<intent-filter>
        <action android:name="android.intent.action.SEARCH" />
</intent-filter>
<meta-data
        android:name="android.app.searchable"
        android:resource="@xml/searchable" />
</activity>
```

2. After associating the search activity with the search action, you must inform Android which activities will have search enabled. You almost always want search available on every activity of your app. In this case, you should define:

```
<application>
...
<meta-data
        android:name="android.app.default_searchable"
        android:value=".RecipeSearch" />
...
</application>
```

This instructs Android to enable search for the entire app using RecipeSearch as the searchable activity. You can also limit search on an activity by activity basis by moving the <meta-data> element directly into individual activities:

```
<activity android:name=".MainMenu">
...
<meta-data
        android:name="android.app.default_searchable"
        android:value=".RecipeSearch" />
...
</activity>
```

This would enable search for MainMenu, and only other activities that have explicitly defined an android.app.default_searchable value for the android:name attribute in <meta-data>.

Build the Search Activity

With the searchable activity defined in the manifest, you can now build the Activity. Users most frequently interact with search results as lists, so it is recommended that you extend ListActivity for your searchable Activity.

The RecipeSearch class should perform three essential functions. First, it should listen for a search query or a search suggestion selection. RecipeSearch will begin loading as soon as the user either executes the query or selects a result from the suggestion list, which means you will have all the information you need at onCreate().

```
@Override
public void onCreate( Bundle savedInstanceState ) {
super.onCreate( savedInstanceState );
setContentView( R.layout.list );
Intent intent = getIntent();
```

```
        if( Intent.ACTION_VIEW.equals( intent.getAction() ) ) {
        //A search suggestion was clicked
            Intent recipeIntent = new Intent( this, RecipeEntry.
➥class );
            recipeIntent.setData( intent.getData() );
            startActivity( recipeIntent );
            finish();
        }
        else if( Intent.ACTION_SEARCH.equals( intent.getAction() ) ) {
        //The search was executed
            String query = intent.getStringExtra( SearchManager.
➥QUERY );
            showRecipes( query );
        }
    }
```

This is the searchable Activity, so you know something about the Intents that can cause this Activity to load—namely that the user has just interacted with a search dialog. By instancing an Intent object by calling getIntent(), you can evaluate the Intent for the next action to take.

If the Intent Action was ACTION_VIEW, you know the user clicked on a single suggestion from the search dialog. You can then launch an Activity to load that result. If ACTION_SEARCH was called, you need to perform the search and display the results.

ANDROID DOES

The Search Manager handles the entire life cycle of the search dialog, but it does not perform the actual search within your app. For that, you need to define your own search methods and handle your own results.

Second, you must execute the search. It is up to you to determine how flexible this search should be. Some of this depends on how you have stored your data. In Chapter 17, you learn to store and retrieve data with databases. From Chapter 8, you know how to store information in SharedPreferences.

Perhaps the most obvious search term is recipe name, in the context of the recipes app. The searchable Activity receives the search query the user entered as a String:

```
        private static final ArrayList<String> _RecipeSearchResults =
➥new ArrayList<String>();
        private ListView resultsView;
```

continues

```
showRecipes( query ) {
        SharedPreferences recipeNames = getSharedPreferences(
➡MainMenu.RecipeNamesPref, RecipeEntry.MODE_WORLD_READABLE);
        String[] recipeList = recipeNames.getString( MainMenu.
➡RecipeNamesPref, null ).split( "," );
        for( String recipe: recipeList )
        {
                if( recipe.contains( query ) )
                RecipeSearchResults.add( recipe );                }

    resultsView = (ListView) findViewById( android.R.id.list
➡);
        resultsView.setAdapter( new ListViewAdapter( this ) );
        resultsView.setTextFilterEnabled( true );
        resultsView.setOnItemClickListener( this );
}
```

This block of code should be familiar from Chapter 8, as it takes the same logic used to generate the list of stored recipes and adds a simple evaluation: if any recipe name contains the search query, this recipe is added to the ListView.

In fact, you can use the RecipesList Activity to function as both your searchable Activity and your navigation Activity if you wanted to consolidate your code base.

Nevertheless, our third goal is to display the results. From this point, the necessary code to inflate the list is identical to the code you wrote for RecipeList—another reason to consider consolidating the code.

If you choose to optimize your code in this way, it is important to understand how Android works with a multipurpose Activity of this kind. Suppose a user were viewing the RecipeList Activity and then tapped the search button. By default, Android would create a second instance of the RecipeList Activity.

In most cases, this behavior is not preferred. Managing only single instances of your activities is much easier. To guarantee this, define the <activity> tag android:launchMode attribute as "singleTop".

```
<activity
android:name=".RecipeList"
...
    android:launchMode="singleTop" >
```

When an Intent receives the ACTION_SEARCH action—a result of the user clicking the search button—it will want to create a new instance of the searchable Activity.

Having set the Activity's launch mode to singeTop, you will want to override this behavior by overriding the onNewIntent() method.

```
@Override
protected void onNewIntent( Intent i ) {
        setIntent( i );
        intentReceived( i );
}
```

You can then reposition your onCreate() intent logic into your new intentReceived() method:

```
intentReceived( Intent newIntent ){
        if( Intent.ACTION_VIEW.equals( newIntent.getAction() )
) {
                //A search suggestion was clicked
                Intent recipeIntent = new Intent( this,
RecipeEntry.class );
                recipeIntent.setData( intent.getData() );
                startActivity( recipeIntent );
                finish();
        }
        else if( Intent.ACTION_SEARCH.equals( newIntent.
getAction() ) ) {
                //The search was executed
                String query = intent.getStringExtra( SearchManager.
QUERY );
                showRecipes( query );
        }
}
```

Providing the Search Option

While many devices feature a dedicated Search button, the feature is strongly recommended but not required. Therefore, you need to implement an alternate mechanism to launch a search. This can be a search icon placed within the page, a button from your main page, or an action available on your options menu.

However you choose to place the search activator, the control should call the onSearchRequested() method. You can also override the method within activities if you want to pause a task pending the results of a search.

When a search is requested, the Activity from which the search was called receives the onPause() event. If the user cancels the search, the previous Activity is restored and onResume() occurs.

Voice Search

One of the holy grails of human and computer interaction has been the ability to communicate in natural, human language and through speech. Voice search on Android allows users to simply speak their query, and Android will automatically transcribe the speech and execute the search. Android began expanding voice search to include multiple languages in Android OS 2.1. Implementing voice search within your app is simple, but it has some limitations.

* Support for non-English languages is still developing.

* Users tend to phrase voice search queries differently. You should make sure your app responds to voice search naturally.

* Voice search queries execute immediately when recording is complete.

To add voice search to your app, modify the searchable.xml resource with the following attributes.

1. Set android:voiceSearchMode to "showVoiceSearchButton". This also requires either the launchWebSearch or launchRecognizer flag. While launchWebSearch will direct the user out of your app to the results of a web search, launchRecognizer will process the speech internally. The result will look like:

```
android:voiceSearchMode="showVoiceSearchButton |
launchRecognizer "
```

2. Define the language model that voice search will support using the android:voiceLanguageModel attribute. This supports two possible values: free_form, which is better optimized for English, and web_search, which is better suited for other languages.

```
android:voiceLanguageModel="free_form"
```

3. Define an optional language override with the android:voiceLanguage attribute. Voice search uses the system default language unless this attribute is defined, which is useful for developers of multilingual apps.

```
android:voiceLanguage="en"
```

Once android:voiceSearchMode is defined, a microphone icon displays next to the search icon in the Quick Search Box. That is it—you have successfully enabled voice search in your app.

More Search Options

Android supports two more key features in the scope of search-recent search queries and custom search suggestions, which you may consider implementing in the future. Each requires a better storage solution, such as a database, which you learn about in Chapter 17 and a ContentProvider, which is discussed in Chapter 18.

Android attempts to optimize the performance of the Quick Search Box by predicting your search terms. It does this by saving the search terms you execute, saving the suggestions you click instead of executing the search, and providing custom search suggestions from each app in the system that has enabled custom search.

You can leverage these techniques to improve the quality of a user's search experience within your own app.

 GOOGLE IT

The Android Developer site provides in-depth guides for implementing advanced search behaviors at http://developer.android.com/guide/topics/search/index.html.

Providing integrated search functionality can dramatically improve the usefulness of your app to your users. Of course, you do not have access to Google's search algorithms, so you must write your own queries and provide your own results. This will take patience and practice to perfect.

The Least You Need to Know

- You must define a searchable Activity and config in the AndroidManifest.xml.
- The Search Manager handles the life cycle of the search dialog and automatically pauses and resumes your activities.
- You must perform your own searches and process their results.
- Voice search enables users to speak their searches.

From Widgets to the Browser

10

In This Chapter

- Creating dialog inputs
- Working with dates and times
- Exploring widgets and controls
- Opening the browser

Well-designed operating systems provide access to a consistent look and feel between screens and settings. This continuity is a benefit both to developers and users. For users, it means that all dialog boxes in Android share a common aesthetic. This makes it easy for users to quickly pick up new apps and start using them with little to no instruction.

For developers, this is a boon. Rather than create date pickers and ratings widgets from scratch, Android has already implemented them for you to use. This chapter explores how to take advantage of these built-in controls from the Date Picker to the Browser.

Widgets and Dialogs

As you build the various activities of your app, you will have occasion to think about the kinds of information you want users to be able to enter and how your app will process that data. *Widgets* and *dialogs* can offer a simple way to solicit a specific kind of response from a user.

You can think of dialogs as a response to some user action. If you have implemented a "Delete All" button, you might respond to the onClick() method with a dialog asking "Are You Sure You Want to Delete All Records?" with "Yes" and "No" buttons.

Dialogs can be used in this way to verify that a user really intends to click a particular control. Dialogs also provide a way to take focus away from your current activity in order to answer a specific question. Dialogs are customizable, but Android supports the following preconfigured dialog types:

- **Dialog.** The base class for all types of dialogs. Presents a configurable dialog message box.

- **AlertDialog.** The most common dialog type. Displays a message with zero to three buttons, a spinner, check boxes, or radio controls.

- **DatePickerDialog.** A dialog that displays a DatePicker widget for selecting a date. See Figure 10.1.

Figure 10.1: *The date setting option that appears when the DatePickerDialog is called.*

- **TimePickerDialog.** A dialog that displays a TimePicker widget for selecting a time. See Figure 10.2.

Figure 10.2: *The time setting option that appears when the TimePickerDialog is called.*

- **ProgressDialog.** A dialog that displays a progress status bar. See Figure 10.3.

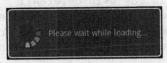

Figure 10.3: *The loading screen that appears when ProgressDialog is called.*

Constructing a Dialog

Dialogs are always created from the context of an Activity, but they are not members of the Activity. Communication between the dialog and the Activity is enabled through callback methods. The dialog life cycle flows through a series of methods and callbacks:

- **showDialog().** Called to begin displaying a dialog.
- **onCreateDialog().** Callback in response to showDialog(). This method is called only the first time the dialog is displayed.
- **onPrepareDialog().** Callback in response to showDialog(), which executes before the dialog is rendered. This method is called every time the dialog is displayed.

On creating the dialog, you will define how your app responds to user input inside the dialog; but what happens if the user does not respond or dismisses the dialog by clicking Back? You can handle either situation with these methods:

- **dismissDialog().** Call to stop showing the dialog. A cached version of the last state of the dialog will be used if the dialog is called again.

- **removeDialog().** Call to remove the dialog's cache and force the dialog to be recreated if called again.

- **onDismiss().** Callback method on the DialogInterface.OnDismissListener interface. Called when a user dismisses the dialog.

You will create the dialog in Java, most commonly in response to an event, such as the onClick() method. To begin, define a simple button in your layout:

```
<LinearLayout>
...
<Button android:id="@+id/deleteAll""
        android:layout_width="wrap_content"
        android:layout_height="wrap_content"
        android:text="Delete All Recipes"
/>
...
</LinearLayout>
```

Next, add the button and an onClickListener to your class's onCreate() method:

```
deleteAllBtn = (Button) findViewById( R.id.deleteAll );
deleteAllBtn.setOnClickListener( new View.OnClickListener() {
        public void onClick(View v) {
                doDialogDisplay()
        }
});
```

Finally, define the **doDialogDisplay()** method and show the dialog:

```
private void doDialogDisplay() {
        AlertDialog.Builder _Builder = new AlertDialog.Builder(
this );
        Builder.setMessage("Are you sure you want to delete all
recipes?")
        .setPositiveButton( "Yes", new DialogInterface.
OnClickListener() {
                public void onClick( DialogInterface dialog, int id ) {
```

```
                    //Execute the delete action
                }
        } )
        .setNegativeButton( "No", new DialogInterface.
OnClickListener() {
                public void onClick( DialogInterface dialog, int id ) {
                        dialog.cancel();
                }
        });
        AlertDialog alert = _Builder.create();
    }
```

Alert dialogs can hold only one control type at a time. If your alert uses buttons, you can have up to three but only one of each kind: positive, negative, and neutral.

- **setTitle().** Defines a title for the alert dialog.

- **setMessage().** Defines the alert message.

- **setCancellable().** Determines whether a user can cancel the dialog with the Back button.

- **setButton().** Defines a button in the dialog.

- **create().** Returns the completed dialog.

Your new alert dialog will cancel the dialog if the user clicks "No". If the user clicks "Yes", you can define an action to purge the recipes from the app.

Dates and Times

In Chapter 16, you will create a ProgressDialog as you learn to work with background threads. The other primary dialog types are for selecting date and time values. There is an entire class of software bug dedicated to issues with mishandled date and time values, which is one of the reasons why developers like to work with native date and time controls.

Using the DatePickerDialog and the TimePickerDialog, you are guaranteed to receive a valid date/time value. You could allow date/time entries in a text box, but this would require validating the value, handling errors, notifying the user to reenter the value, and converting it to and from String types. With the native dialogs, all of this work is neatly handled for you.

You can now modify the RecipeEntry activity to include a recipe "Created Date" field, which will consist of a TextView for displaying a read-only date and an ImageButton to activate the DatePickerDialog and select and set the date.

```
<LinearLayout>
...
    <TextView
            android:id="@+id/createdDate"
            android:layout_width="wrap_content"
            android:layout_height="wrap_content"
            android:text=""
    />
    <ImageButton
            android:id="@+id/datePicker"
            android:layout_width="wrap_content"
            android:layout_height="wrap_content"
            android:src="@drawable/calendar_icon"
    />
    ...
</LinearLayout>
```

The DatePickerDialog will need a unique integer to identify it. Declare this in your Activity class:

```
static final int DATE_PICKER = 0;
```

Additionally, the DatePickerDialog requires a month, day, and year to initialize. Declare properties to store these values:

```
private int _Year;
private int _Month;
private int _Day;
private Calendar _Calendar;
private Button pickDate;
private TextView createdDate;
```

Set the date values to the current year, month, and day and assign the onClickListener for the button inside onCreate():

```
protected void onCreate( Bundle savedInstanceState ) {
    //...
    final Calendar _Calendar = Calendar.getInstance();
    Year = _Calendar.get( Calendar.YEAR );
    Month = _Calendar.get( Calendar.MONTH );
    Day = _Calendar.get( Calendar.DAY_OF_MONTH );
```

```
                createdDate = (TextView)findViewById( R.id.createdDate );
                pickDate = (Button)findViewById( R.id.datePicker );
                pickDate.setOnClickListener( new View.OnClickListener()
    {
                public void onClick( View v ) {
                        showDialog( DATE_PICKER );
                }
            } );
        }
```

Once showDialog() is called, onCreateDialog() and onPrepareDialog() will immediately follow. Override these methods in order to define the DatePicker dialog:

```
    @Override
        protected Dialog onCreateDialog( int i ) {
        switch( i ) {
        case DATE_PICKER:
        return new DatePickerDialog( this,
                    DateSetListener,
                    _Year, _Month, _Day );
            default:
            return null;
            }
        }
```

The DatePickerDialog correctly launches when the user clicks the calendar icon for the ImageButton, but you might want to do something with the final selected date value. For this, you need to define the callback for the onDateSet() method:

```
        private DatePickerDialog.OnDateSetListener mDateSetListener =
    new DatePickerDialog.OnDateSetListener() {
                public void onDateSet( DatePicker view, int year,
                int month, int day ) {
                    _Year = year;
                    _Month = month;
                    _Day = day;
                    setCreatedDate();
                }
            };
```

The year, month, and day value have now been set with the values the user selected in the dialog. You can now define the setCreatedDate() method to update the createdDate TextView object with the new date value:

```
        private void setCreatedDate()  {
            StringBuilder date = new StringBuilder();
```

continues

```
                        date.append(_Month + 1).append("/")
                        .append(_Day).append("/")
                        .append(_Year);
                        createdDate.setText( date.toString() );
        }
```

This completes the DatePickerDialog, which is now ready to test with your app. In normal use, you will probably want to do more with the date than simply display it on screen. You will likely want to save the value for permanent record and possibly validate the value. Additionally, the TimePickerDialog follows the same process, though you would use hour and minute values instead of year, month, and day.

More Widgets and Controls

You have already worked with a variety of widgets, including controls like Buttons, EditTexts, and Spinners as well as View elements like ListViews, LinearLayouts, and TabHosts. Android offers many more specialized widgets which provide unique control possibilities.

You have already worked with DatePicker, which the DatePickerDialog implements for you inside the dialog box. You can implement this directly inside your Activity's layout, although as you have seen from the size of the widget, it takes up quite a bit of space on screen.

For this reason, some widgets are better implemented in dialogs, while others fit nicely on screen. Android sets no aesthetic restrictions on how you design your app, so experiment with each way. Some of the more detailed widgets include:

- **AnalogClock.** Displays an analog clock with hands for minutes and hours. See Figure 10.4.

Figure 10.4: *The analog clock widget.*

Here's the code for the clock:

```
<AnalogClock
        android:id="@+id/AnalogClock"
        android:layout_width="wrap_content"
        android:layout_height="wrap_content">
</AnalogClock>
```

- **Chronometer.** Displays a simple timer, which can be used to count up from a time or count down to a time.

The code for the chronometer is:

```
<Chronometer
        android:text="@+id/Chronometer"
        android:id="@+id/Chronometer"
        android:layout_width="wrap_content"
        android:layout_height="wrap_content">
</Chronometer>
DatePicker. Displays a date picker.
<DatePicker
        android:id="@+id/DatePicker"
        android:layout_width="wrap_content"
        android:layout_height="wrap_content">
</DatePicker>
```

- **DigitalClock.** Displays a digital clock. See Figure 10.5.

The digital clock has a brief code:

```
<DigitalClock
        android:text="@+id/DigitalClock"
        android:id="@+id/DigitalClock"
        android:layout_width="wrap_content"
        android:layout_height="wrap_content">
</DigitalClock>
```

Figure 10.5: *The digital clock widget.*

- **RatingBar.** Shows a rating as a configurable number of stars. See Figure 10.6.

Figure 10.6: *The rating bar widget.*

The rating bar also has brief code:

```
<RatingBar
    android:id="@+id/RatingBar01"
    android:layout_width="wrap_content"
    android:layout_height="wrap_content">
</RatingBar>
```

- **SlidingDrawer.** A special widget which allows content to be hidden off-screen and dragged on-screen. This requires two child views—one for the handle which is dragged, the second for the hidden content. See Figure 10.7.

Figure 10.7: *The sliding drawer widget exposes more content when touched.*

The sliding drawer has a longer code because it needs extensive text:

```
<SlidingDrawer
        android:id="@+id/SlidingDrawer"
        android:layout_width="wrap_content"
        android:layout_height="wrap_content"
        android:handle="@+id/handle"
        android:content="@+id/content">
        <TextView
            android:id="@id/handle"
            android:layout_width="wrap_content"
            android:layout_height="wrap_content"
            android:text="Slide Me to Display More"
        />
        <TextView
            android:id="@id/content"
            android:layout_width="match_parent"
            android:layout_height="match_parent"
            android:text="More Content to View Here"
        />
</SlidingDrawer>
```

• **TimePicker.** Displays a time picker. See Figure 10.8.

Figure 10.8: *The time picker widget.*

The time picker code is short and sweet:

```
<TimePicker
        android:id="@+id/TimePicker01"
        android:layout_width="wrap_content"
        android:layout_height="wrap_content">
</TimePicker>
```

These widgets are fully functional after their simple XML definitions, but you can fully customize each of them through either their XML attributes or from within your Java code. Many of the more advanced Android features actually correspond to simple widget classes which you define with a few lines of XML. For example, multimedia viewers hook into a simple MediaController widget, which you will learn about in Chapter 11.

Integrating the Browser

It is easy to forget sometimes that before the explosion of app storefronts, the key feature of modern smartphones was the rich, mobile browser. Android's browser uses the *WebKit* rendering engine to display websites and pages.

DEFINITION

WebKit is an open source layout engine that allows browsers to render content. WebKit powers Google's Chrome and Android mobile browsers as well as Apple's Safari and Mobile Safari for the iPhone.

Working with browsers in Android, you have two primary options—you can open a new instance of the Android browser or you can embed the browser as a widget inside your app. There are a variety of reasons to include a browser in your app which are supported by Android:

- Open the default Android browser based on a link stored in your app.

- Embed a browser to open your app's own local HTML files.

- Embed a browser to open and interact with specific sites.

- Embed a browser to extend or change the default browsing experience.

Opening the Web Page

Frequently, you do not want to invest the time and overhead to incorporate new infrastructure directly into your app. Suppose you wanted to add a way to quickly e-mail recipe information to a user's contacts. While you could build your own e-mail client into your app to do this, it would be much faster to simply allow Android to launch its own e-mail client and handle all of that work for you.

In a similar fashion, you probably do not want to manually manage the browser instance for every possible site that your app might collect—it is more efficient and easier to simply allow Android to launch a new browser instance in a separate Activity based on the links you collect.

It is a straightforward process with Android. For this example, create an EditText and a button in your layout. You will enter the website in the text box and click the button to open the site in a browser, as shown in Figure 10.9.

Figure 10.9: *Opening up the browser in your app.*

```
import android.app.Activity;
import android.content.Intent;
import android.net.Uri;
import android.os.Bundle;
import android.view.View;
import android.view.View.OnClickListener;
import android.widget.Button;
import android.widget.EditText;

public class OpenBrowserActivity extends Activity{        Button
  openSite;
      EditText siteURL;

      @Override
      public void onCreate( Bundle savedInstanceState ) {
             super.onCreate( savedInstanceState );
             setContentView( R.layout.main );

             siteURL = (EditText)findViewById( R.id.textURL );
             openSite = (Button)findViewById( R.id.buttonOpen );
             openSite.setOnClickListener( new OnClickListener() {
                    public void onClick( View v ) {
                           launch();
                    }
             });
      }

      private void launch() {
             Uri u = Uri.parse( siteURL.getText().toString() );
             startActivity( new Intent(Intent.ACTION_VIEW, u) );
      }
}
```

As you can see, the launch() method requires only two lines of code to open the browser with the entered website.

In practice, you will probably allow users to store websites with their recipes, which they will later click to launch the browser.

PITFALL

For websites, the URI must include the "http://" prefix in order for Android to interpret the Intent request. You should validate the URL string before attempting to parse it as a URI.

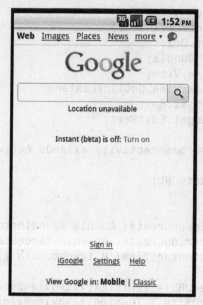

Figure 10.10: *The Google launch page from within your app.*

Linkify

As if this process were not easy enough, Android provides a helper class aptly named Linkify, which instantly converts all of the content of any view into hyperlinks that call an Intent.ACTION_VIEW with the correct URI.

```
TextView text: URL = (TextView) findViewById( R.id.textUrl );
textURL.setText( "http://www.google.com" );
Linkify.addLinks( textURL , Linkify.ALL );
```

Linkify has one primary method, addLinks(), which can transform the text of any View into actionable links. You can even define your own regular expressions to convert text into links customized for your app. Linkify supports the following types of links as constants:

- **ALL.** Suggests that all matching patterns be converted to links.

- **EMAIL_ADDRESSES.** Converts only e-mail addresses matching format: name@text.tld. Links will open the default mail app.

- **MAP_ADDRESSES.** Converts only matching street addresses. Links will open the Google Maps app at the matching address.

- **PHONE_NUMBERS.** Converts only phone numbers. Links will open the Phone app for the matching number.

- **WEB_URLS.** Converts only matching URLs. Links will open the default browser at the matching website.

GOOGLE IT

The Android SDK documents how to use regular expressions to define your own custom link actions using your own ContentProvider. Visit the WikiNotes sample app tutorial to get started.

Embedding the Browser

While opening a new browser instance is quick and easy to implement, it has the disadvantage of directing users out of your app. If you plan to allow your app to interact with certain websites or if you want to create a more seamless app experience, you may wish to embed a browser inside your app.

Android has simplified this process with a unique WebView widget. With a few extra lines of code, you can modify the OpenBrowserActivity Activity class you created previously to embed a browser in a WebView widget that renders in the same screen.

First, add the WebView widget to the Activity layout with the text box and launch button (see Figure 10.11):

```xml
<?xml version="1.0" encoding="utf-8"?>
<LinearLayout
        xmlns:android="http://schemas.android.com/apk/res/android"
        android:layout_width="fill_parent"
        android:layout_height="fill_parent"
        android:orientation="vertical"
>
        <LinearLayout
                android:layout_width="fill_parent"
                android:layout_height="wrap_content"
                android:orientation="horizontal"
        >
                <EditText
                        android:text="Enter a website"
                        android:id="@+id/textURL"
                        android:layout_width="wrap_content"
                        android:layout_height="wrap_content">
```

continues

```
                </EditText>
                <Button
                        android:text="Open"
                        android:layout_height="wrap_content"
                        android:id="@+id/buttonOpen"
                        android:layout_width="wrap_content">
                </Button>
        </LinearLayout>
        <WebView
                android:id="@+id/Browser"
                android:layout_height="wrap_content"
                android:layout_width="wrap_content">
        </WebView>
</LinearLayout>
```

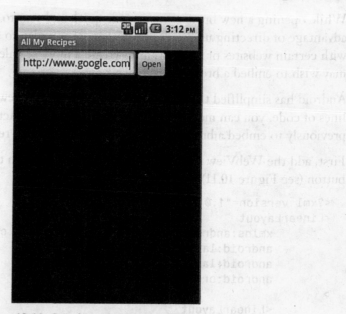

Figure 10.11: *Searching the web from within the app.*

You will now see the same Activity screen as before, except that the browser will now be embedded within the view.

With only a few modifications to your OpenBrowserActivity class, you will have a fully managed WebView (see Figure 10.12):

```
import android.app.Activity;
import android.os.Bundle;
```

```
import android.view.View;
import android.view.View.OnClickListener;
import android.widget.Button;
import android.widget.EditText;
import android.webkit.WebView;
import android.webkit.WebViewClient;
public class OpenBrowserActivity extends Activity{

        Button openSite;
        EditText siteURL;
        WebView browser;

        @Override
        public void onCreate( Bundle savedInstanceState ) {
                super.onCreate( savedInstanceState );
                setContentView( R.layout.web );

                siteURL = (EditText)findViewById( R.id.textURL );
                openSite = (Button)findViewById( R.id.buttonOpen );
                browser = (WebView)findViewById( R.id.Browser );
                browser.setWebViewClient( new BrowserWebViewClient() );

                openSite.setOnClickListener( new OnClickListener() {
                        public void onClick( View v ) {
                                launch();
                        }
                });
        }
    private void launch() {
        //Uri u = Uri.parse( siteURL.getText().toString() );
        //startActivity( new Intent(Intent.ACTION_VIEW, u) );
        browser.getSettings().setJavaScriptEnabled( true );
        browser.loadUrl( siteURL.getText().toString() );
    }
    private class BrowserWebViewClient extends WebViewClient {
        @Override
        public boolean shouldOverrideUrlLoading( WebView view, String
url ) {
                view.loadUrl( url );
                return true;
                }
        }
    }
```

Figure 10.12: *Browsing Google embedded in your app.*

Whenever using Android classes, which allow open access to the Internet, remember to declare permission in the AndroidManifest.xml:

```
<uses-permission android:name="android.permission.INTERNET" />
```

The two key changes to your OpenBrowserActivity class are that you have included a WebView widget in your layout XML and instanced it in Java, and that you have called the WebView.loadUrl() method to load a site inside your embedded browser.

It is also good practice to override default URL handling by creating your own WebViewClient class. This ensures that all browsing activity inside your WebView will stay in your WebView and not launch a new instance of the default Android browser.

ANDROID DOES

You can interact with pages in your WebViews directly by using JavaScript. This enables you to communicate information directly between your app and a remote website or vice versa. Call loadUrl() with "javascript:" followed by the exact code to run.

The Least You Need to Know

- All visible elements within an app correspond to Widget classes.
- Most widgets can be defined entirely in your XML layout resources.
- Dialogs can separate key workflow input from your primary Activity screens.
- Use the Android browser to launch websites or embed them within your app.

Make the Most of the Hardware

The Android platform has penetrated phones, tablets, and even TVs, so the hardware specs and advantages are wide ranging. Part 3 shows you how to make the most of their features, including cameras, 3D visuals, and location-based services.

Make the Most of
the Hardware

Cameras and Media

In This Chapter

- Using the camera to take a photo
- Recording audio and video
- Working with multimedia files
- Embedding media playback

Can you remember the last mobile phone you bought that did not include a camera? Even the free with contract phones from most carriers include cameras, multimedia apps to play audio and video, and photo gallery apps to browse and share photos.

Most current and next generation Android phones come with powerful High Definition (HD) cameras, HD displays, and a suite of powerful multimedia apps to take advantage of both. This chapter guides you through the basics of implementing the camera, working with photos, and using audio and video media to enhance your app.

Introduction to Multimedia on Android

The Android SDK implements support for a variety of hardware and software features that enable apps to both capture input and display stored and captured media. Every Android device is different, but the SDK supports all of these features:

- Camera with auto-focus, zoom, color balance, and flash
- Video recording from camera
- Audio recording from microphone

- Photo gallery manager
- Audio and video playback with MediaPlayer and the Android JET media engine

PITFALL

While the Android SDK supports all of these features, you cannot assume that specific features, like auto-focus or zoom, exist on all devices. Write your app code to check for features first before attempting to use one that might not exist on a user's device.

Many of these features require *permission* that must be declared in AndroidManifest. xml using the syntax:

```
<uses-permission android:name="android.permission.CAMERA" />
```

DEFINITION

Permission is the clearance that many Android features require from the user, like the device's current GPS location. Permissions are defined in the manifest, and the user is prompted to accept or deny permission to the features after the app is installed.

All of the multimedia functionality in Android is encapsulated in a few, concise classes.

- The **Camera** class provides all the resources necessary to take pictures and switch to video recording.
- The **MediaPlayer** class provides a way to play audio and video files and streams.
- The **VideoView** class allows video playback inside a View.
- **MediaRecorder** class offers a way to record audio and video.
- The **SoundPool** class provides a way to play back dynamic sound effects.

While the Android emulator can simulate playback from most of these classes, the emulator cannot simulate actual multimedia input. This means that the Camera and MediaRecorder classes will be quite limited in an emulated environment. In order to fully test the features you build with these classes, you will need a physical Android device.

Incorporating the Camera into Your App

Before smartphones, cameras did little else than take pictures. These pictures were prisoners on a roll of film or memory card until removed for processing and development by some other tool. Android devices are powerful computers that happen to take pictures.

Google Goggles takes images from the camera to translate the text into different languages, find the photographed object on the Internet, and even locate it on a map. Evernote takes photos and transcribes text into searchable notes. In Android, the camera can be used to do much more than simply collect photos. Think about the camera as a versatile tool that can add significant creative value to your apps.

Prepare the SurfaceView

Thankfully, Android devices do not have viewfinders. The camera input preview is rendered on screen, nested in a View. The Camera class requires a SurfaceView layout in order to render, which is quite simple to define in XML:

```
<LinearLayout  xmlns:android="http://schemas.android.com/apk/res/
  android"
        android:layout_width="fill_parent"
        android:layout_height="fill_parent"
        android:screenOrientation="landscape"
        >
        <SurfaceView
        android:id="@+id/surface_camera"
        android:layout_width="fill_parent"
        android:layout_height="fill_parent"
        android:layout_weight="1"
        >
        </SurfaceView>
</LinearLayout>
```

Android automatically manages the camera preview inside the defined SurfaceView space. This View can also hold child controls, such as buttons, if you need to implement specialized behavior. Do not overload the SurfaceView, however, because rendering the camera preview is an expensive operation.

The Camera class itself requires permission, which must be declared in AndroidManifest.xml and approved by the user when the app is installed. To begin, you will want to request the following:

```
<uses-permission android:name="android.permission.CAMERA" />
```

It is good practice to use the <uses-feature> element to indicate hardware and software features that your application uses. If you do not specify any features, Android assumes by the permission that you want all available features included. The android:required attribute indicates whether the feature is required for the app to function. The default value is true, so it is best to explicitly define your app's requirements, as shown here:

```
<uses-feature android:name="android.hardware.camera"
    android:required="true"/>
<uses-feature android:name="android.hardware.camera.autofocus"
    android:required="false"/>
```

GOOGLE IT

The Android Developer site provides a complete list of hardware and software features which should be declared in the manifest. Visit http://developer. android.com/guide/topics/manifest/uses-feature-element.html for the complete list.

To start working with the camera immediately, you need only to call the ACTION_IMAGE_CAPTURE intent by attaching it to a method, such as a button's onClick():

```
Intent newPictureIntent = new Intent( android.provider.MediaStore.
    ACTION_IMAGE_CAPTURE );
startActivityForResult( newPictureIntent, requestCode );
```

Android automatically selects the camera for the task and returns the result of the image capture, which can be received in the onActivityResult() method, which can be processed by:

```
protected void onActivityResult(int requestCode, int resultCode,
    Intent data)
{
        Bitmap newPictureBmp = (Bitmap) data.getExtras().get( "data"
    );
}
```

This produces a quick response and return from the camera.

Working the Camera Class

The Camera class offers much more sophisticated control options for the camera. The simplest way to do this is by creating a new CameraView Activity class to tailor the camera to your needs. The basic steps for interacting with the camera by method begin:

1. Instance a Camera object and call **open()**.
2. Fetch the existing or default settings by calling **getParameters()**.
3. If desired, alter the **Camera.Parameters** return object and call **setParameters(Camera.Parameters)**.
4. Call **setDisplayOrientation(int)** if needed.
5. Give **setPreviewDisplay(SurfaceHolder)** a **SurfaceHolder** object. In order to display a preview, a surface is required.
6. Prior to capturing content, **startPreview()** must be called to start updating the preview surface.
7. Call the appropriate methods to capture either photos or video.
8. Call **startPreview()** again to capture more media. The preview stops after each capture.
9. Stop updating the preview by calling **stopPreview()**.
10. Release the camera as soon as it is no longer needed by **release()**.

PITFALL

The camera uses significant resources. Minimize your apps impact by releasing the camera immediately in onPause(). Reopen the camera in onResume().

For example, here is a basic Camera class:

```
public class CamaraView extends Activity implements SurfaceHolder.
    ➡Callback {

        private SurfaceView mySurfaceView;
        private SurfaceHolder mySurfaceHolder;
        private Camera myCamera;
```

continues

```
        private boolean isPreviewRunning;

        @Override
        public void onCreate( Bundle savedInstanceState ) {
                super.onCreate( savedInstanceState );
                setContentView( R.layout.camera_view );
                mySurfaceView = ( SurfaceView ) findViewById( R.id.
surface_camera );
                mySurfaceHolder = mySurfaceView.getHolder();
                mySurfaceHolder.addCallback( this );
                //This instructs the surface not to use its own buffers
                mySurfaceHolder.setType( SurfaceHolder.SURFACE_TYPE_
PUSH_BUFFERS );
        }

    public void surfaceCreated( SurfaceHolder holder ) {
                myCamera = Camera.open();
                if( camera != null ){
                        Camera.Parameters params = camera.
getParameters();
                        camera.setParameters( params );
                }
                else {
                        Toast.makeText( getApplicationContext(), "There
is no camera available.", Toast.LENGTH_LONG ).show();
                        finish();
                }
    }


    //surfaceCreated() is a change and will trigger this method
    public void surfaceChanged( SurfaceHolder holder, int format,
int w, int h ) {
                if( isPreviewRunning ) {
                        myCamera.stopPreview();
                }
                Camera.Parameters p = myCamera.getParameters();
                p.setPreviewSize( width, height );
                p.setPreviewFormat( PixelFormat.JPEG );
                myCamera.setParameters(p);
                try {
                        myCamera.setPreviewDisplay( holder );
                        myCamera.startPreview();
                        isPreviewRunning = true;
                }
                catch (IOException e) {
```

```
            e.printStackTrace();
      }
}

//This method is called at onPause() or onDestroy()
public void surfaceDestroyed( SurfaceHolder holder ) {
      myCamera.stopPreview();
      isPreviewRunning = false;
      myCamera.release();
}

}
```

This initial class structure has prepared the activity to capture content—either photos or video. You need to implement additional methods to perform the capture.

Capturing the Photo

With the CameraView class in place, you have everything you need to prepare the camera to take photos. Decide where to store these images, what format to store them in, and whether they should be public or private for your app. The takePicture() method on Camera class performs the work of capturing an image through the camera.

takePicture has four parameters set up like takePicture: Camera.ShutterCallback shutter, Camera.PictureCallback raw, Camera.PictureCallback postview, Camera. PictureCallback jpeg.

Here are the details on the takePicture() data:

- **Shutter.** A callback for the moment of image capture. Can be null.

- **Raw.** A callback for the uncompressed (raw) image data. Can be null.

- **Postview.** A callback for the postview image data. Can be null.

- **Jpeg.** A callback for the JPEG image data. Can be null.

Here you can process a user's photos by adding a private _takePicture() method to your activity, which stores the image on external storage in the Pictures directory:

```
private void _takePicture() {
      myCamera.takePicture(null, null, myPhotoCallback);
```

continues

```
        }

        Camera.PictureCallback myPhotoCallback = new Camera.
➡PictureCallback() {
            public void onPictureTaken( byte[] data, Camera camera
➡) {
                new storeNewImage().execute( data );
                camera.startPreview();
            }
        };
        class storeNewImage extends AsyncTask<byte[], String, String>
{
            protected String doInBackground(byte[]... jpg) {
                File photo = new File( Environment.DIRECTORY_
➡PICTURES,"photo.jpg" );
                if (photo.exists()) {
                    photo.delete();
                }
                try {
                    FileOutputStream fos = new
➡FileOutputStream( photo.getPath() );
                    fos.write( jpg[0] );
                    fos.close();
                }
                catch ( java.io.IOException e ) {
                    e.printStackTrace();
                }
                return(null);
            }
        }
    }
```

Recording Video

Using the same CameraView class, you can implement video recording. You will also need to request permission to record audio, if you want both video and audio together.

```
<uses-permission android:name="android.permission.RECORD_AUDIO" />
```

Use the initial steps for interacting with the camera; when ready to record video, the following additional steps are needed:

1. Allow the media process access to the camera by calling **unlock()**.

2. Give a camera object to **setCamera(myCamera)**.

3. At the end of recording, call **reconnect()** to find and relock the camera.

4. Restart the preview in order to take more photos or videos.

Using these startRecording() and stopRecording() methods, your app can capture video and store it in the external Pictures directory.

```
private MediaRecorder myMediaRecorder;
private final int maxDurationInMs = 20000;
private final long maxFileSizeInBytes = 500000;
private final int videoFramesPerSecond = 20;
public void startRecording(){
    try {
        myCamera.unlock();
        myMediaRecorder = new MediaRecorder();
        myMediaRecorder.setCamera( myCamera );
        myMediaRecorder.setAudioSource( MediaRecorder.
➥AudioSource.MIC );
        myMediaRecorder.setVideoSource( MediaRecorder.
➥VideoSource.CAMERA );
        myMediaRecorder.setOutputFormat( MediaRecorder.
➥OutputFormat.MPEG_4 );
        myMediaRecorder.setMaxDuration(maxDurationInMs
➥);
        File videoFile = new File( Environment.
➥DIRECTORY_PICTURES,"video.mp4" );
        myMediaRecorder.setOutputFile( videoFile.
➥getPath() );
        myMediaRecorder.setVideoFrameRate(
➥videoFramesPerSecond );
        myMediaRecorder.setVideoSize( mySurfaceView.
➥getWidth(), mySurfaceView.getHeight() );
        myMediaRecorder.setAudioEncoder( MediaRecorder.
➥AudioEncoder.DEFAULT );
        myMediaRecorder.setVideoEncoder( MediaRecorder.
➥VideoEncoder.DEFAULT );
        myMediaRecorder.setPreviewDisplay(
➥mySurfaceHolder.getSurface() );
        myMediaRecorder.setMaxFileSize(
➥maxFileSizeInBytes );
        myMediaRecorder.prepare();
        myMediaRecorder.start();
    }
    catch (IllegalStateException e) {
```

continues

```
                              e.printStackTrace();
            }
            catch (IOException e) {
                          e.printStackTrace();
            }
      }

      public void stopRecording(){
            myMediaRecorder.stop();
            myCamera.lock();
      }
```

The most important and unfortunately least well-documented aspect of video recording on Android is that all required properties must be set. Your app compiles and runs, but when attempting to record, you may see unhelpful errors like "prepare failed." Prior to calling prepare(), be sure that you have given all properties appropriate values.

Accessing Stored Photos

Photo and video files are stored just as any other file in the filesystem. When working with media directly from the camera, you will probably use explicit or relative paths and filenames to place the content in the appropriate folders. While you could navigate the filesystem to find media, Android provides a MediaStore class which provides all types of media from both internal and external storage.

To present a gallery of all photos on the device, launch an ACTION_GET_CONTENT or an ACTION_PICK intent. This will display a gallery-like list of the media types you specify. Do this by associating an intent with a method on a relevant button or widget:

```
Intent mediaChooser = new Intent( Intent.ACTION_GET_CONTENT );
mediaChooser.setType( "images/*" );
startActivityForResult( mediaChooser ,1 );
```

Once the user selects something from the gallery, you will want to do something with that information. Override the onActivityResult() method to specify some action to take with the returned data.

```
@Override
      public void onActivityResult( int requestCode, int resultCode,
   ➡Intent data ) {
```

```
                        super.onActivityResult( requestCode, resultCode, data
    );
                if( requestCode == 1 ) {
                        if( resultCode == Activity.RESULT_OK ) {
                                Uri selectedPhotoLocation = data.
    getData();
                                // Do something with the data
                        }
                }
        }
```

Embedding Audio or Video

Android supports native playback of a variety of media formats; the most recognizable of these are probably MP3 for audio and MP4 for video. You have the option of including your own media resources in the /res/raw directory, using multimedia available from the MediaStore or by streaming media from the web.

Regardless of where the multimedia content comes from, playing these audio or video sources in your app is straightforward. The MediaPlayer class can serve all of your multimedia playback needs.

To embed a raw resource into an activity, instance a MediaPlayer object and call start():

```
MediaPlayer myPlayer = MediaPlayer.create( context, R.raw.crickets_
    chirping );
    myPlayer.start();
```

MediaPlayer's methods are:

- **create().** Creates an instance of MediaPlayer from a raw resource. Calls **prepare()** the first time.

- **start().** Begins playing the media file.

- **pause().** Pauses media playback. Can be resumed.

- **stop().** Stops media playback. Cannot be resumed.

- **reset().** Resets media playback. It automatically calls to **prepare()** after **stop()** is called.

- **prepare().** Prepares media for playback. Must call before **start()** after **reset()**.

Embedding a MediaPlayer using another data source is equally simple:

```
MediaPlayer myPlayer = new MediaPlayer();
if( null != myPlayer ) {        myPlayer.setDataSource( path_to_a_file_
→or_webstream );
    myPlayer.prepare();
    myPlayer.start();
}
```

PITFALL

Because any app can embed media, it is possible for a MediaPlayer object to return null on declaration. It is good practice to ensure the object exists before attempting to use it.

Android supports a parallel process for displaying video content through VideoView, a subclass of SurfaceView. The VideoView class allows for scaling and tinting of video files and functions similarly to MediaPlayer. Define a <ViewGroup> in your XML layout, and implement it in an activity:

```
VideoView videoView = (VideoView) findViewById( R.id.VideoView );
MediaController myMediaController = new MediaController( this );
myMediaController.setAnchorView( videoView );
Uri video = Uri.parse( web_link_to_video );
videoView.setMediaController( myMediaController );
videoView.setVideoURI( video );
videoView.start();
```

ANDROID DOES

Android can play almost any content from the web through a MediaPlayer or MediaController object, as long as the media is capable of progressive downloading.

Using the Microphone

Audio recording follows the same steps as video recording, without the video handling. To record audio-only content, follow these steps:

1. Instance a new **MediaRecorder** object.
2. Instance a new **ContentValues** object, and assign standard properties like **TITLE**, **TIMESTAMP**, and **MIME_TYPE**.

3. Identify a place to store the content or use a **ContentResolver** to assign a path automatically from the **Content** database.

4. Call **setAudioSource()** to define the source.

5. Call **setOutputFormat()** to define the output format.

6. Call **setAudioEncoder()** to define the encoder.

7. Call **prepare()** to ready the **MediaRecorder** object.

8. Call **start()**.

9. Call **stop()**.

10. **Release()** it.

Once ready, you can implement audio recording in your app.

```
MediaRecorder myAudioRecorder = new MediaRecorder();
ContentValues myContentValues = new ContentValues(3);

myContentValues.put( MediaStore.MediaColumns.TITLE, "Notes for
➥My App" );
myContentValues.put( MediaStore.MediaColumns.TIMESTAMP,
➥System.currentTimeMillis() );
myContentValues.put( MediaStore.MediaColumns.MIME_TYPE,
➥myAdioRecorder.getMimeContentType() );

ContentResolver myContentResolver = new ContentResolver();

Uri base = MediaStore.Audio.INTERNAL_CONTENT_URI;
Uri newUri = myContentResolver.insert( base, values );

String path = myContentResolver.getDataFilePath( newUri );

myAudioRecorder.setAudioSource( MediaRecorder.AudioSource.MIC
➥);
myAudioRecorder.setOutputFormat( MediaRecorder.OutputFormat.
➥MP3);
myAudioRecorder.setAudioEncoder( MediaRecorder.AudioEncoder.
➥AMR_NB );
myAudioRecorder.setOutputFile( path );

myAudioRecorder.prepare();
myAudioRecorder.start();
```

The Least You Need to Know

- The Camera class provides everything you need to interact with the physical camera.
- To playback media files, use the MediaPlayer.
- VideoViews provide some additional configuration options for video playback.
- MediaRecorder will record both audio and video content.
- You must request permission to use many dedicated hardware resources in Android.

Location-Based Services

In This Chapter

- Accessing the device's location context
- Work with system service listeners
- Listening to Android's sensors
- Using Google Maps

Location awareness has become perhaps one of the more subtle paradigm shifts in consumer technology. As almost every new mobile device includes GPS, Wi-Fi location fetching, and communication tower triangulation, services from Google Latitude, FourSquare, and Facebook Places allow users to instantly check-in, showing their location and the location of their friends.

MapQuest and Google Maps provide free, turn-by-turn directions with voice navigation. With an Android device in hand, you can instantly know where you are, what direction you are moving, and where friends and family happen to be at the same time.

Location services in Android allow you to add context to your app and its data. With location, you can adjust baking instructions for altitude or find the closest grocery store to buy a missing ingredient. This chapter introduces location receivers and sensors, allowing you to make your app aware of its surroundings.

Introduction to Android Sensors and Receivers

Many of the most powerful assets of an Android device are hidden inside the case, visible only from the code in your app. The display, buttons, and even the camera are visible hardware features, but much more waits beneath the surface waiting to be used.

Most Android devices include a range of hardware features that fall into two categories: *receivers* and *sensors*. Receivers include:

> **DEFINITION**
>
> **Receivers** process information from transmitting stations, such as antennas, towers, and satellites. In mobile devices this communication typically occurs over Radio Frequency (RF).
>
> **Sensors** return information collected from the device itself, such as touch screen input, rotation, and movement.

- **Wi-Fi radio.** Connects to Wi-Fi Internet hotspots and receives location information from public Wi-Fi databases.

- **Carrier antenna.** Sends and receives calls. Receives location information by tower triangulation.

- **GPS.** The Global Positioning System (GPS) receiver returns location information collected from satellites.

Sensors provide much more specific types of information about the current state and position of the device relative to its environment. Sensors include:

- **Proximity.** Detects the distance between the device and another object. Frequently used to dim the display as it approaches the ear.

- **Light.** Detects changes in light. Often used to brighten or darken the display in response to available light.

- **Front-facing camera.** Can be used for facial recognition. Sometimes used with location tracking of stolen devices.

- **Accelerometer.** Tracks changes in the speed of the device's movement. Can be used to assist a pedometer.

- **Compass.** Calculates device's orientation to north.

- **Gyroscope.** Returns the device's physical orientation by three axes.

- **Temperature.** Detects the temperature of the environment surrounding the device.

- **Pressure.** Returns the current atmospheric pressure at the device's location.

Just as with the camera, the available hardware features vary by device and platform. Phones and tablets are likely to have GPS receivers, whereas Google TV devices are not. Generally, Android anticipates possible gaps in hardware and allows for software feature calls to degrade gracefully.

ANDROID DOES

In many cases, the Android SDK assumes that certain hardware features do not exist on a particular device. With a location information request, Android can attempt to try GPS first, Wi-Fi position next, tower triangulation, and Internet address location last. In this way, some information always returns even if not the most accurate.

Android provides ways to access both explicit sensor information from a particular hardware device and more generic data from the sensors and receivers on the actual device. Think about your target audience and optimize your code for the features you must have, leaving room for the users without these hardware features.

Location, Location

The location context of an Android device contains significant information, far more than just the actual coordinates on a map. The quality of this information enables powerful apps to provide driving and walking directions around the world, but it also requires responsibility.

Privacy is always a huge concern for end users, so the first step to begin using location data is to request permission:

```
<uses-permission android:name="android.permission.ACCESS_FINE_
➥LOCATION" />
```

By accepting this, access is granted to both fine GPS location, which is very accurate, and coarse Wi-Fi and tower triangulation location, which is more broad and, therefore, a bit more private. For coarse location only, request:

```
<uses-permission android:name="android.permission.ACCESS_COARSE_
  ➡LOCATION" />
```

PITFALL

Just because users accept permission to share their location when they install your app does not mean that they always want to share their location information. Respect your users' privacy by clearly communicating within your app when you intend to use location services and what you intend to do with the data.

Consider creating a special series of settings in your options menu to turn location services on and off. You can make this as granular as necessary, but as a general rule, options should be as concise as possible while still conveying the scope of the settings.

Location Services

In Android, location requests are handled through the LocationManager class, which provides access to the system services managing location information. Working with this class, you can determine the current location of the device, request periodic updates of location, and send Intents when the device nears a particular location.

Regardless of how you plan to use location information inside your app, the logical flow should follow these steps:

1. In response to an event, begin collecting updates from location providers. Continue to receive updates to maintain.

2. When necessary, stop requesting location updates.

3. Do something with the collected location data.

To get started, you would define a LocationListener to return callback information from a LocationManager instance. You can implement this as an Activity that implements the LocationListener class and renders the location latitude/longitude coordinates in a simple TextView.

```
import android.app.Activity;
import android.location.Criteria;
import android.location.Location;
import android.location.LocationListener;
import android.location.LocationManager;
```

```
import android.location.LocationProvider;
import android.os.Bundle;
import android.widget.TextView;
import android.content.Context;
public class LocationData extends Activity implements
   LocationListener{
        private LocationManager myLocationManager;
        private TextView myLocationText;
        private String myLocationProvider;
        private Location myLocation;
        private Criteria bestProvider;
        @Override
        public void onCreate(Bundle savedInstanceState) {
               super.onCreate(savedInstanceState);
               setContentView(R.layout.about);

               myLocationManager = (LocationManager) this.
   getSystemService(Context.LOCATION_SERVICE);
               myLocationText = (TextView) this.findViewById( R.id.
   about_layout );

               bestProvider = new Criteria();
               myLocationProvider = myLocationManager.getBestProvider(
   bestProvider, true );
               myLocation = myLocationManager.getLastKnownLocation(
   myLocationProvider );
               String lastLocation = Double.toString( myLocation.
   getLatitude() ) + ", " + Double.toString( myLocation.
   getLongitude() );
               myLocationText.setText( lastLocation );
        }
}
```

The LocationManager object provides most of the methods needed to begin working with location data. It first requires a network provider that can be explicitly declared by using LocationManager.GPS_PROVIDER for GPS or LocationManager. NETWORK_PROVIDER for Wi-Fi and tower. Using the getBestProvider() method enables Android to select the highest quality provider at the moment. Outdoors, this is likely GPS, while inside a building it might be a network.

Once you have the LocationManager instance, you can call getLastKnownLocation() to return the last location information Android received and stored. This information could be out of date, so requestLocationUpdates() can be called to begin polling for current location. The method requestLocationUpdates() is *overloaded*, and can be used to call an Intent on location update or to call the onLocationChanged() method of the listener.

DEFINITION

In Java, a method is said to be **overloaded** if multiple methods exist with the same name that require different parameters. Methods that serve similar purposes are sometimes overloaded to improve code readability and reduce code complexity.

To use the method to call an intent on a location update, you need to supply some parameters for requestLocationUpdates(String provider, long minTime, float minDistance, PendingIntent intent):

- **provider.** The location provider to service the updates.
- **minTime.** The minimum time in milliseconds between notifications. This time is a power conservation guideline, as actual execution times will vary.
- **minDistance.** The minimum distance in meters from the last notification location to send the next notification.
- **intent.** An intent to execute on notification.

To use the method to notify the listener that a location has updated and thereby call the onLocationChanged() method:

```
requestLocationUpdates( String provider, long minTime, float
minDistance, LocationListener listener )
```

Here, the parameters are the same except for the last: listener. Because you implemented the LocationListener class, onLocationChanged() becomes a required method of the LocationData activity. In onCreate(), you can now implement:

```
myLocationManager.requestLocationUpdates( myLocationProvider, 2500,
0, this );
```

And later define:

```
public void onLocationChanged( Location location ) {
    // Do something with the new location
}
```

A location object has several useful methods that describe the direction, altitude, accuracy, and coordinates of the location.

Here are location class methods, some of which require another piece of data to function:

Location Class Methods

Method	Parameters	Returns
getAccuracy()	None	The accuracy of the location in meters
getAltitude()	None	The altitude of the location
getBearing()	None	The direction of the location in degrees east of true north
getLatitude()	None	The latitude of the location
getLongitude()	None	The longitude of the location
getSpeed()	None	The speed of movement in meters per second
getTime()	None	The time the location was recorded
distanceTo()	Location	The distance in meters between the current location and parameter
bearingTo()	Location	The direction in degrees east of true north from the current location to the supplied location

The LocationListener class requires three additional methods to respond to possible changes from the LocationManager. You have already implemented onLocation-Changed(), which receives a Location object at the minimum update intervals you specified.

It is also possible that a LocationProvider can become unavailable. If getBestProvider() returns the network provider, and the user switches to Airplane mode, the selected provider becomes disabled. Two methods respond to disabling and enabling a LocationProvider:

```
public void onProviderEnabled( String provider ) {
        //Provider is now available for use
}
public void onProviderDisabled( String provider ) {
        //Provider is no longer available.
}
```

Finally, it is possible that the selected provider can fail to complete a request. If GPS is selected and the user walks into a tall building, the GPS provider will likely return TEMPORARILY_UNAVAILABLE. When the user leaves the building, the provider will probably return AVAILABLE.

If you are using location in your app, you'll want a particular action for when GPS location availability ends or begins:

```
public void onStatusChanged( String provider, int status, Bundle
    extras ) {
        //Provider availability status has changed
}
```

Unless you plan to design a navigation app, you probably do not want to let location update requests continue running. For most purposes, you might want to fetch a user's location once, perhaps in response to clicking a button and then immediately release the system services from your LocationManager object. When finished with location services, call removeUpdates():

```
myLocationManager.removeUpdates( this );
```

If you do need frequent location updates, be sure to anticipate app interruptions and implement onPause() and onResume() events:

```
@Override
protected void onPause() {
        super.onPause();
        myLocationManager.removeUpdates( this );
}
@Override
protected void onResume() {
        super.onResume();
        myLocationManager.requestLocationUpdate(
myLocationProvider, 2500, 0, this );
}
```

Consider Accuracy and the Battery

As a rule, do not execute code unless necessary. Dedicated GPS devices are frequently attached to a power source, and they can poll satellites as many times per second as their processors are capable. Even without considering good programming philosophy, most Android devices spend most of their running time on battery power. When executing power-expensive operations, like requesting location updates, this has not only a cost for the processor and memory, but also the battery.

Consider the following questions when you implement your location code:

- How frequently does your app need location updates from providers? Do not poll more than this number.
- Can you manage your own listening window around other logic in your app to reduce the frequency of updates?
- How accurate do all updates need to be? Network updates are cheaper than GPS updates. Do not use GPS if you only need the nearest city.

While it is important to consider battery life and try to optimize your code to run more efficiently, sometimes you also need to ensure that your information is accurate. All location receivers have margins of error. Weather, temperature, and time of day can affect the quality of your location data.

Depending on the app, this margin of error can mean the difference between geo-tagging a family photo over the drop off of a waterfall or giving driving directions off a cliff. If you need to provide the user with data as accurate as possible, you need to collect and compare location data.

Consider creating a Location Collection that includes all locations returned over a period of time. If you collect ten locations over two minutes, it should be straight-forward to evaluate the accuracy of the locations as well as the consistency. A series of high-accuracy readings in a straight line should probably be favored over low-accuracy readings scattered about.

Emulator Limitations

If you have a physical Android device, you can test your app's location code by taking a short walk down the street. GPS receivers are accurate up to 30 feet with a strong signal, and if you live in an urban area, there are likely to be plenty of recognized Wi-Fi hotspots and towers to triangulate your position. If you do not yet have an Android device, or even if you do, you will still want to be able to test your code for basic errors in the emulator.

The emulator does not attempt to simulate receiver hardware, so it will not return meaningful information for speed, direction, and altitude. In fact, by default, the emulator uses a single, static location as the response for all requests to location services. You can supply alternative location coordinates from within Eclipse and through the Dalvik Debug Monitor service.

Start the emulator, and then select from the Eclipse menu: **Window > Show View > Other... > Android > Emulator Control.** By default, this opens in a new tabbed Eclipse view at the bottom of your workspace. Scroll to the bottom, and you will see the option to send the latitude and longitude coordinates manually, as shown in Figure 12.1.

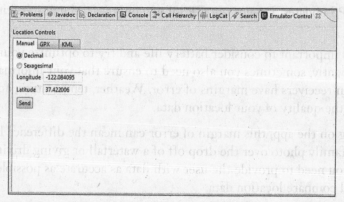

Figure 12.1: *You can manually simulate GPS coordinates in the Eclipse emulator.*

You can also do this in DDMS using the same Emulator Control view. Open the DDMS perspective from **Window > Open Perspective > Other... > DDMS.**

Sense and Sensors

If you have used an Android device, you are probably already familiar with some of the many sensors on the device—even if you have not thought about them individually. You might have noticed that the screen on the device brightens and dims in response to the light in a room or that the screen blanks when holding it next to your ear. If you have used Google's free Sky Map app, you have noticed that the app automatically brings the constellation at which you point the device into focus.

All of this magical behavior is made possible by many different hardware sensors, coordinated by Android system services. Unlike receivers, sensors respond only to variables in the local environment. It would be difficult to construct personally identifiable information based on the angle the device is held and the distance from an ear; therefore, you do not need to ask permission to use these services and features.

In fact, sensors are generally features that users expect apps to implement. The most obvious of these is orientation. Users expect apps to respond to rotating the device between landscape and portrait mode. Android handles some of this work for you, but to make the most of the available sensors, you need to implement them in code.

Reading a Sensor

Just as with location, sensors are managed by system services, and they are instanced in the same way. Here is the code for accessing the SensorManager class:

```
private SensorManager mySensorManager;
mySensorManager = (SensorManager) getSystemService( Context.SENSOR_
  ⇒SERVICE );
```

Also like the LocationManager class, the SensorManager class maintains all sensors on the device. Unlike locations, sensors can communicate vastly different types of information. Surrounding temperature and device acceleration are unrelated bits of information at best. Given this, you cannot simply call getBestSensor() because there are too many types of sensors for a method like that to be meaningful.

Instead, you can call getSensorList() to return a List of sensors on the device matching the type of sensor requested. Alternatively, call getDefaultSensor() to return the sensor Android decides is most suited to the task. Both methods take an integer parameter to define the type of sensor to return. For example, getSensorList(Sensor. TYPE_ALL) would return all sensors on the device. Other available constants include:

- **TYPE_ACCELEROMETER.** Sensors that detect any device momentum on all three axes.

- **TYPE_GYROSCOPE.** Sensors that measure relative movement of the device on all three axes.

- **TYPE_LIGHT.** Sensors that measure changes in light in device's environment.

- **TYPE_MAGNETIC_FIELD.** Sensors that measure changes in magnetic attraction against all three axes.

- **TYPE_PRESSURE.** Sensors that measure pressure applied to the surface of the device, usually against the touchscreen.

- **TYPE_PROXIMITY.** Sensors that measure the distance between the device and another object.

- **TYPE_TEMPERATURE.** Sensors that return some temperature information.

Android also supports orientation of the device to north, and still includes a TYPE_ORIENTATION in the SDK. This type is now deprecated, and Google encourages users to implement specific methods getRotationMatrix() to be used with remapCoordinateSystem() and getOrientation() in order to process orientation information.

PITFALL

At times, the Android OS develops faster than the hardware market. Pressure and temperature sensors are not consistently implemented, if at all, in current devices. While light, acceleration, proximity, and orientation sensors are widely implemented and supported, other sensors referenced in the Android SDK may vary in implementation between devices.

Sensor services in Android can be implemented into an Activity class in a similar manner as location services. It is a three-step process:

1. Create an activity that implements the SensorEventListenor class.

2. Call registerListener() to begin receiving change notifications in the onSensorChanged() method.

3. Call the unregisterListener() method either when finished or onPause() to maintain efficiency.

```
import java.util.List;
import android.app.Activity;
import android.hardware.Sensor;
import android.hardware.SensorEvent;
import android.hardware.SensorEventListener;
import android.hardware.SensorManager;
import android.os.Bundle;
import android.content.Context;
public class LocationData extends Activity implements
    ➡SensorEventListener{
```

```java
        private SensorManager mySensorManager;

        @Override
        public void onCreate(Bundle savedInstanceState) {
                super.onCreate(savedInstanceState);
                setContentView(R.layout.about);

                mySensorManager = (SensorManager) getSystemService(
        Context.SENSOR_SERVICE );

                List<Sensor> mySensors = mySensorManager.getSensorList(
        Sensor.TYPE_ACCELEROMETER );
                for(Sensor s: mySensors)
                {
                        mySensorManager.registerListener( this, s,
        SensorManager.SENSOR_DELAY_NORMAL );
                }
        }
        public void onAccuracyChanged( Sensor arg0, int arg1 ) {
                // Respond to changes in Accuracy

        }
        public void onSensorChanged( SensorEvent event ) {
                // Respond to new sensor data
        }
        @Override
        public void onResume() {
                super.onResume();
                Sensor mySensor = mySensorManager.getDefaultSensor(
        Sensor.TYPE_ACCELEROMETER );
                mySensorManager.registerListener( this, mySensor,
        SensorManager.SENSOR_DELAY_NORMAL );
        }

        @Override
        public void onPause(){
                super.onPause();
                mySensorManager.unregisterListener( this );
        }
}
```

Sensor Data

Sensors do not have to wait on external receivers to communicate data. They process information much faster, because the data is immediately available. Therefore, sensors can more quickly return much more information than location providers. Different Android devices process sensor data differently, so it is important to test your app on as many different devices as possible.

The onSensorChanged() method receives a SensorEvent object. Like a Location object, the SensorEvent contains all of the information you need to work with the data including properties for accuracy, time, the sensor itself, and the data values.

The values property is an array of floating point values. The contents and their meaning vary by sensor type. You can assume that the array positions will always be the same, but be sure to review the SensorEvent class at the Android developer site with each new release of the Android SDK.

- **TYPE_ACCELEROMETER.** Values measure the acceleration applied to the phone minus the force of gravity.
  - values[0]: Acceleration minus Gx on the x-axis
  - values[1]: Acceleration minus Gy on the y-axis
  - values[2]: Acceleration minus Gz on the z-axis
- **TYPE_MAGNETIC_FIELD.** Values measure the ambient magnetic field on the x, y, and z axes.
- **TYPE_LIGHT.** Single value, values[0], represents the ambient light level.
- **TYPE_PROXIMITY.** Single value, values[0] representing distance in centimeters. Some devices report only a binary value for "near" or "far."

GOOGLE IT

The OpenIntents group provides an Android sensor test utility which mimics various Android sensors by using the computer's mouse. Their SensorSimulator project is hosted at http://code.google.com/p/openintents/wiki/SensorSimulator.

Google did not attempt to implement any sensor simulation into the Android emulator, so you will not be able to test any of your sensor code without third-party software tools or a physical device. The available software tools can certainly eliminate computational bugs in your code, but you will not be able to escape buying at least a few different Android devices to fully test a Sensor app.

Integrate Maps

If you collect location data, what do you want to do with a user's latitude and longitude coordinates? Some developers like to write their own code to use these values in unique ways for their apps. The rest of us probably want something very simple: show the location on a map.

To do this, you can instance a browser in your app and let the web apps for Google Maps, Bing Maps, or MapQuest do the work of rendering the page. Google provides a simpler way that requires some additional tools and registration, but will make mapping much easier.

Register for Google Maps Access

Unlike the Android source code, the Google Maps source code is not open. The data and the APIs are free for anyone to use, but they require a separate license agreement between you as a developer and Google. This requires two steps. First, generate a certificate to sign your app, and second, request an API key from Google with the certificate.

GOOGLE IT

The complete licensing agreement and requirements for working with the Google Maps API are hosted at http://code.google.com/android/add-ons/google-apis/maps-api-signup.html.

Additionally, you need to install the Google APIs add-on in the Android SDK Manager. See Figure 12.2.

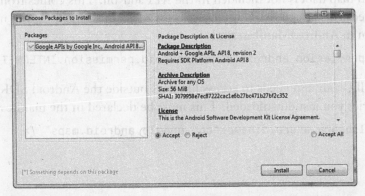

Figure 12.2: *Adding the Google APIs from the Android SDK Manager.*

These were probably installed when you first set up Eclipse in Chapter 1, but you can double check by doing the following:

1. Launch the Android SDK Manager in your SDK application folder.

2. If the Google APIs for API 8 (Android 2.2) are not installed, you will see a window asking you to accept the agreement and install them.

3. Accept the terms and click **Install.**

4. Once the download completes, you are ready to go.

Working with Google Maps

With the Google APIs installed, you can now implement Google's maps into your app. Most of the work of using maps involves setting up your Eclipse environment to use the new APIs. Follow these steps to do that:

1. Either create a new Android project in Eclipse or change your current project to target the Google APIs.

2. Right-click the project, select **Properties > Android > Google APIs** for the most current platform (Platform 2.2, as of this writing).

3. Create a new Android emulator for the target build.

4. Select **Window > Android SDK and AVD Manager > New....**

5. Set the Target to Google APIs (Google Inc.)— and the most current API (API Level 8, as of this writing).

The actual map data is not included in the API add-on. This comes from Google's servers. Because the app needs Internet access to acquire this data, you must request permission in AndroidManifext.xml:

```
<uses-permission android:name="android.permission.INTERNET" />
```

Additionally, your app needs to access libraries outside the Android SDK, namely, the API libraries you just downloaded. This must be declared in the manifest as well:

```
<uses-library android:name="com.google.android.maps" />
```

The key class in the Google API is MapView, which is a subclass of ViewGroup. This means that you define a map layout resource that implements a MapView and includes your Google Maps API key:

```
<com.google.android.maps.MapView
        android:id="+id/mapview"
        android:apiKey="Your API Key"
        android:clickable="true"
        android:layout_height="fill_parent"
        android:layout_width="fill_parent"
/>
```

The MapView is part of an external library, which means that it must be defined by its full package path: com.google.android.maps.MapView. Use your own API key to enable the map's functionality. Creating an activity to run your map is just as simple:

```
import com.example.android.google.apis.R;
import com.google.android.maps.MapActivity;
import android.os.Bundle;
public class myMapView extends MapActivity {
    @Override
    public void onCreate(Bundle savedInstanceState) {
            super.onCreate(savedInstanceState);
            setContentView(R.layout.map);
    }
    @Override
    protected boolean isRouteDisplayed() {
            return false;
    }
}
```

You can actually run this activity right now in the emulator. An unfocused map will appear, which you can pan around. In order to add more specific controls to the view, you can expand onCreate() to include:

```
MapView myMapView = (MapView) findViewById( R.id.mapview );
myMapView.setBuiltInZoomControls( true );
```

The MapActivity class handles the entire life cycle of the Activity, fetches data from the Internet, and cleans up after itself. To begin, the only thing you need to do with a MapView is add some content to it, such as a marker for the current location of a user. This content is added to the map as an overlay using the MyLocationOverlay class.

Adding content to a MapView involves these steps:

1. Create a **MapController** instance. The **MapController** uses calls **animateTo()** to set the starting location and **setZoom()** to zoom the map to a specific level.

2. Create a **MyLocationOverlay** instance. The **MyLocationOverlay** class provides methods **enableMyLocation(),** which instance the necessary location providers and listeners; **enableCompass(),** which automatically activates the orientation sensors; and **runOnFirstFix(),** which defines the behavior upon the first location fix received.

3. Add the overlays to the **MapView** by using **getOverlays().add().**

You can further extend **onCreate():**

```
final MapController myController = myMapView.getController();
final MyLocationOverlay myOverlay = new MyLocationOverlay(
this, myMapView );
myOverlay.enableMyLocation();
myOverlay.enableCompass();
myOverlay.runOnFirstFix( new Runnable(){
        public void run(){
                myController.setZoom( 7 );
            myController.animateTo( myOverlay.getMyLocation() );
            }
});
myMapView.getOverlays().add( myOverlay );
```

This provides the basic framework for working with Google's map infrastructure. If you needed to collect information based on what a user selects in a map, the API TrackballGestureDetector class provides several methods to return map and grid information based on what the user taps or selects.

The Least You Need to Know

- You must define permissions to use location services and Google Maps.
- Preserve battery life and maximize performance by limiting location and sensor listeners running time.
- Google Maps integration requires agreeing to their API terms and conditions.

- Location services provide fine and course location information about the device.
- Sensor services provide information about the device's environment.

3D Graphics and Animation

In This Chapter

- Introducing OpenGL ES
- Understanding key terms and classes
- Creating a surface for 3D
- Modeling an object in 3D

Almost two decades ago, films like *Jurassic Park* had just begun to popularize the use of three-dimensional, computer-generated images—a technique that required numerous supercomputers to complete the animation. Most modern smartphones now have the graphics hardware to render effects and animation that just recently required "super" computers.

This chapter introduces the software and hardware rendering capabilities of Android devices. You learn the key terminology and syntax for creating three-dimensional objects and animation.

Introduction to 3D Graphics

While 3D televisions, games, and films have recently resurged in popularity, 3D in the context of graphic design does not mean that the images are holographic or require special glasses. Instead, 3D graphics simply create the illusion of depth. Shadows, wrinkles, and contours can all add depth of field, making an image or graphic seem to occupy a real space.

Android implements 3D graphics through the OpenGL (Open Graphics Library), which was first developed by Silicon Graphics. OpenGL provides a platform-neutral

API for rendering 2D and 3D graphics, mainly targeting desktop and game systems. The Khronos Group manages a subset of OpenGL known as OpenGL ES (OpenGL for Embedded Systems), which is optimized for mobile devices. As of Android OS 2.2, OpenGL ES 1.0 is fully implemented, and many OpenGL ES 2.0 APIs have been included.

Although OpenGL ES is optimized for mobile devices, advanced 3D rendering still requires a processor capable of drawing complex graphics. As with other Android hardware, you cannot assume that all devices are created equal. Many devices without high-performance processors will be able to render some level of detail and quality, so it is good practice to make the complexity of your graphics configurable by the user.

OpenGL Basics

Many apps require few if any fancy graphics and can use Android's built-in animations to look and feel polished and professional. On the other hand, apps like Google Earth depend on three dimensionalities in order to deliver the core app experience. Even if you do not need 3D in your app today, you should know what Android is capable of delivering.

If you have developed 3D models before, you will already be comfortable with some of the core terminology. Otherwise, familiarize yourself with the following key terms:

- **Vertex.** A vertex is a single point in three-dimensional space and is the starting reference point for many objects. A vertex can have as few as two coordinates (X,Y) and as many as four (X,Y,Z,W). If omitted, the W axis has a default value of 1.0 and the Z axis has a value of 0. Unless otherwise noted, a point and vertex are synonymous.

- **Triangle.** A triangle is composed of three points or vertices.

- **Polygon.** A polygon is an object that has at least three connected points. A triangle is also a polygon.

- **Primitives.** Primitives are 3D objects composed of either triangles or polygons.

OpenGL rendering depends on two core classes. First, GLSurfaceView, which is an implementation of the SurfaceView class. This is the view upon which the object is drawn. Android decouples the thread for the OpenGL object, which means the GLSurfaceView requires explicit onPause() and onResume() life cycle methods. The

View also requires a renderer. As the name suggests, the renderer draws or renders the objects.

Second, you need a renderer that implements the GLSurfaceView.Renderer interface. This class requires three methods:

- **onSurfaceCreated().** Called when the OpenGL Surface is created. Set values which are least likely to change in this method.

- **onDrawFrame().** The actual drawing occurs in this method.

- **onSurfaceChanged().** The surface creation is the first change and will call this method once. Changes to the size of the Surface will call this method, most obviously rotating the device between portrait and landscape mode.

These methods each take a GL10 class object, which you can think of as the object drawn or rendered. This class supports some key methods worth reviewing:

- **glEnable().** Enables various rendering features. All except GL_DITHER and GL_MULTISAMPLE are off by default.

- **glDisable().** Disables rendering features.

- **glClearColor().** Sets the color of the clipping wall, or the furthest visible distance, beyond which all other items are invisible.

- **glClear().** Clears the current color settings to reveal color changes.

- **glClearDepthf().** Sets the depth of the buffer.

- **glHint().** Performs some perspective calculations.

The GL10 class supports many rendering settings, expressed as constants, which can be enabled by glEnable(). Some of these include:

- **GL_BLEND.** Mixes new color values with the colors already in the buffer.

- **GL_DITHER.** Increases the number of perceived colors by juxtaposing contrasting colors.

- **GL_LIGHTING.** Enables light and darkness calculations.

- **GL_MULTISAMPLE.** Performs antialiasing (smoothing) on the edges of polygons, improving perceived image quality.

- **GL_POINT_SMOOTH.** Performs antialiasing on points.

Drawing an Activity

No matter how small or large your graphic vision may be, like almost everything else in Android, OpenGL graphics begins with an Activity and a View. In this case, you create the View objects in Java—no XML required. To simplify the process, each component is compartmentalized in its own class. In this way, a single class handles the Activity, another the rendering, and another the object definition.

Begin by creating the Activity:

```
import android.app.Activity;
import android.os.Bundle;
public class OpenGLActivity extends Activity {
    private OpenGLView _OpenGLView;

    @Override
    protected void onCreate( Bundle savedInstanceState )          {
        super.onCreate( savedInstanceState );
        OpenGLView = new OpenGLView( this );
        setContentView( _OpenGLView );
    }

    @Override
    protected void onPause() {
        super.onPause();
    }
    @Override
    protected void onResume() {
        super.onResume();
    }
}
```

PITFALL

Remember to define onPause() and onResume() events for OpenGL activities. Doing so greatly improves the user experience, and more importantly prevents stability issues for your app.

As OpenGLView does not yet exist, you need to create the class, which should extend GLSurfaceView:

```
import java.util.Random;
import android.content.Context;
```

```
import android.opengl.GLSurfaceView;
import android.view.MotionEvent;
public class OpenGLView extends GLSurfaceView {
        private OpenGLRenderer _OpenGLRenderer;
        public OpenGLView( Context context ) {
                super( context );
                OpenGLRenderer = new OpenGLRenderer();
                setRenderer( _OpenGLRenderer );
        }
        public void onTouchEvent(final MotionEvent event) {
                queueEvent(new Runnable() {
                        public void run() {
                                Random x = new Random();
                                renderer.setColor(x.nextFloat(),
➥x.nextFloat(), x.nextFloat()));
                        }
                });
        }
}
```

With these two classes in place, you have an Activity and a View to display on screen. All that remains is to render some content inside that View, which you define OpenGLRenderer to accomplish:

```
import javax.microedition.khronos.egl.EGLConfig;
import javax.microedition.khronos.opengles.GL10;
import android.opengl.GLSurfaceView;
public class OpenGLRenderer implements GLSurfaceView.Renderer {
        private float _R = 0.2f;
        private float _G = 0.4f;
        private float _B = 0.6f;

        @Override
        public void onSurfaceCreated( GL10 gl, EGLConfig config ) {
                // Nothing yet
        }
        @Override
        public void onSurfaceChanged( GL10 gl, int w, int h ) {
                gl.glViewport( 0, 0, w, h);
        }
        @Override
        public void onDrawFrame( GL10 gl ) {
                gl.glClearColor( _R, _G, _B, 1.0f );
                gl.glClear( GL10.GL_COLOR_BUFFER_BIT );
```

continues

```
    }
    public void setColor( float r, float g, float b ) {
    R = r;
    G = g;
    B = b;
    }
}
```

You can run this app now in the emulator. You will see a solid color background that changes to a new random color with every click.

Model Objects

Much of your 3D modeling will likely take place in a dedicated design application, where you can create the environments, models, and textures. You then import these objects into your Android project for rendering through OpenGL. In the meantime, you can render simple shapes to work to experience some of the possibilities of 3D design in Android.

GOOGLE IT

Many development resources and engines have already been built to help you rapidly develop 3D environments. OpenIntents references a number of free and open source projects at http://www.openintents.org/en/libraries. Subscription offerings such as Unity3 are also available at http://unity3d.com. Blender and AndCAD also offer Android compatible 3D modeling solutions.

The Cube

The API demos in the Android SDK provide a great way to get started with working with 3D animations. When you have finished with this chapter, you can create a new Eclipse project targeting Android OS 2.0.1 using the sample ApiDemos. Explore the OpenGL classes for functional and interactive models. For now, we will create the framework for a cube. See Figure 13.1.

Figure 13.1: *The 3D cube we are creating.*

Create a new Cube class to store the definition. As you see in the code, the vertices[], colors[], and indices[] arrays are used to define their respective buffers:

```
import java.nio.ByteBuffer;
import java.nio.ByteOrder;
import java.nio.IntBuffer;
import java.util.Random;
import javax.microedition.khronos.opengles.GL10;
class Cube
{
        private IntBuffer _VertexBuffer;
        private IntBuffer _ColorBuffer;
        private ByteBuffer _IndexBuffer;

        public Cube()
            {
            int one = 0x10000;
            int vertices[] = {
            one, -one, -one,
            one, -one, -one,
            one,  one, -one,
            one,  one, -one,
            one, -one,  one,
```

continues

```
            one, -one,  one,
            one,  one,  one,
            one,  one,  one,
    };
    int colors[] = {
            0,     0,    0,  one,
            one,   0,    0,  one,
            one,  one,   0,  one,
            0,  one,     0,  one,
            0,    0,  one,  one,
            one,   0,  one,  one,
            one,  one,  one,  one,
            0,  one,  one,  one,
    };
    byte indices[] = {
            0, 4, 5,    0, 5, 1,
            1, 5, 6,    1, 6, 2,
            2, 6, 7,    2, 7, 3,
            3, 7, 4,    3, 4, 0,
            4, 7, 6,    4, 6, 5,
            3, 0, 1,    3, 1, 2
    };
    ByteBuffer vbb = ByteBuffer.allocateDirect( vertices.length*4
);
    vbb.order( ByteOrder.nativeOrder() );
    VertexBuffer = vbb.asIntBuffer();
    VertexBuffer.put( vertices );
    VertexBuffer.position( 0 );
    ByteBuffer cbb = ByteBuffer.allocateDirect( colors.length*4 );
    cbb.order( ByteOrder.nativeOrder() );
    ColorBuffer = cbb.asIntBuffer();
    ColorBuffer.put( colors );
    ColorBuffer.position( 0 );
    IndexBuffer = ByteBuffer.allocateDirect( indices.length );
    IndexBuffer.put( indices );
    IndexBuffer.position( 0 );
}
```

Because you can manipulate all of these aspects of the object later, it is most important that you define the way the object should look when it first appears. Everything else can be changed later.

Buffer objects are used to store, initialize, and render vertex arrays and element indexes from memory. By assigning the arrays into the buffers, the vertex, color, and index, definition of the cube is ready to retrieve instantly.

```
public void draw( GL10 gl )
{
        gl.glFrontFace( GL10.GL_CW );
        gl.glVertexPointer( 3, GL10.GL_FIXED, 0, _VertexBuffer
);
        gl.glColorPointer( 4, GL10.GL_FIXED, 0, _ColorBuffer );
        gl.glDrawElements( GL10.GL_TRIANGLES, 36, GL10.GL_
UNSIGNED_BYTE, _IndexBuffer );
}
}
```

By creating a single, simple draw() method on the Cube class, you can immediately instance a new cube object as needed in your app.

Moving the Cube in 3D

The Cube class by itself will not do anything; you must instance it elsewhere to render and manipulate it. Just as before, you will need an Activity, a GLSurfaceView, and a GLSurfaceView.Renderer class. This time, you can create them all within the same parent class, CubeRotate:

```
import javax.microedition.khronos.egl.EGLConfig;
import javax.microedition.khronos.opengles.GL10;
import android.app.Activity;
import android.content.Context;
import android.opengl.GLSurfaceView;
import android.os.Bundle;
import android.view.MotionEvent;
public class CubeRotate extends Activity {
        private GLSurfaceView _GLSurfaceView;

        @Override
        protected void onCreate( Bundle savedInstanceState )          {
                super.onCreate( savedInstanceState );
                GLSurfaceView = new CubeSurfaceView( this );
                setContentView( _GLSurfaceView );
                GLSurfaceView.requestFocus();
                GLSurfaceView.setFocusableInTouchMode( true );
        }
        @Override
```

continues

```
        protected void onResume() {
                super.onResume();
                GLSurfaceView.onResume();
        }
        @Override
        protected void onPause() {
                super.onPause();
                GLSurfaceView.onPause();
        }
}
```

So far, this should be familiar ground. The CubeRotate activity will instance a new CubeSurfaceView, which you will create as:

```
class CubeSurfaceView extends GLSurfaceView {
private final float TOUCH_SCALE_FACTOR = 180.0f / 320;
private final float TRACKBALL_SCALE_FACTOR = 36.0f;
private CubeRenderer _CubeRenderer;
private float _LastX;
private float _LastY;

public CubeSurfaceView( Context context ) {
        super( context );
        CubeRenderer = new CubeRenderer();
        setRenderer( _CubeRenderer );
        setRenderMode( GLSurfaceView.RENDERMODE_WHEN_DIRTY );
}
```

You will create the GLSurfaceView.Renderer, CubeRenderer, last. By calling using RENDERMODE_WHEN_DIRTY on setRenderMode(), this instructs Android to render only when the Surface is rendered or when requestRender() is called.

```
@Override
public boolean onTrackballEvent( MotionEvent e ) {
        CubeRenderer.renAngleX += e.getX() * TRACKBALL_SCALE_FACTOR;
        CubeRenderer.renAngleY += e.getY() * TRACKBALL_SCALE_FACTOR;
        requestRender();
        return true;
}
```

Later in the class, you will make the cube rotate and change color. By listening for onTrackballEvent(), you capture the X and Y angle of the trackball motion and use that to re-render the object.

```
@Override
public boolean onTouchEvent( MotionEvent e ) {
        float x = e.getX();
        float y = e.getY();

        switch ( e.getAction() ) {
            case MotionEvent.ACTION_MOVE:
                float dx = x - _LastX;
                float dy = y - _LastY;
                CubeRenderer.renAngleX += dx * TOUCH_SCALE_FACTOR;
                CubeRenderer.renAngleY += dy * TOUCH_SCALE_FACTOR;
                requestRender();
        }
        LastX = x;
        LastY = y;
        return true;
}
```

Similarly, the cube will rotate in response to touch, but in this case with more precision as the user will drag the cube into the desired position. In this case, you will need both the direction the cube is moving as well as the final coordinates (when the user has released the cube).

```
private class CubeRenderer implements GLSurfaceView.Renderer {

        private Cube Cube;
        public float renAngleX;
        public float renAngleY;

        public CubeRenderer() {
            Cube = new Cube();
        }
        public void onSurfaceCreated( GL10 gl, EGLConfig config ) {
            gl.glDisable( GL10.GL_DITHER );
            gl.glHint( GL10.GL_PERSPECTIVE_CORRECTION_HINT, GL10.
    ➡GL_FASTEST );
            gl.glClearColor( 1,1,1,1 );
            gl.glEnable( GL10.GL_CULL_FACE );
            gl.glShadeModel( GL10.GL_SMOOTH );
            gl.glEnable( GL10.GL_DEPTH_TEST );
        }

        public void onSurfaceChanged( GL10 gl, int width, int height )
    ➡{
        gl.glViewport( 0, 0, width, height );
```

continues

```
        float ratio = (float) width / height;
        gl.glMatrixMode( GL10.GL_PROJECTION );
        gl.glLoadIdentity();
        gl.glFrustumf( -ratio, ratio, -1, 1, 1, 10 );
    }
```

Notice that, inside the GLSurfaceView.Renderer object, the actual cube will be drawn. In onSurfaceCreated(), the first life cycle method to launch, GL_DITHER is disabled to improve performance. The clipping wall is set to white with glClearColor(). The additional methods enable GL_CULL_FACE, which instructs the rendered to ignore back polygons and GL_DEPTH_TEST, which calls for depth comparisons.

Because onSurfaceCreated() is a change, onSurfaceChanged() executes next. Here the cube is rendered as a two-dimensional square, where it waits for interaction from the user to animate.

```
public void onDrawFrame(GL10 gl) {
        gl.glClear( GL10.GL_COLOR_BUFFER_BIT | GL10.GL_DEPTH_BUFFER_
    BIT );
        gl.glMatrixMode( GL10.GL_MODELVIEW );
        gl.glLoadIdentity();
        gl.glTranslatef( 0, 0, -3.0f );
        gl.glRotatef( renAngleX, 0, 1, 0 );
        gl.glRotatef( renAngleY, 1, 0, 0 );
        gl.glEnableClientState( GL10.GL_VERTEX_ARRAY );
        gl.glEnableClientState( GL10.GL_COLOR_ARRAY );
        Cube.draw(gl);
    }
```

Last, but most importantly, the cube's color buffers are set and Cube.draw() is called to render the cube. When you run the app, at first the cube will appear to be a multi-colored but 2D image. As soon as you touch the cube, you will see it render in 3D with colors changing as the cube moves. See Figure 13.2.

Figure 13.2: *The 3D cube being manipulated in the Android emulator.*

The Least You Need to Know

- 3D graphics in Android are rendered through OpenGL ES.
- GLSurfaceView and GLSurfaceView.Renderer are the two core classes needed to implement 3D models.
- Android renders 3D objects in a separate thread, which is separate from your application's threads.
- All 3D objects in OpenGL are composed of vertices and primitives.
- The ApiDemos in the Android SDK are a great way to learn more about OpenGL.

Core Services

In This Chapter

- Listening to the phone service
- Managing network connections
- Using Wi-Fi services
- Connecting with Bluetooth

Although many Android device owners think of their phones and tablets as portable computers—computers that also happen to make calls and send messages—communication capability is perhaps the most important feature of any mobile device. Whether by voice, text message, e-mail, or chat, communication is integral to the Android experience.

All Android devices connect to a communication channel: Google TVs directly connect to the Internet, Android tablets connect to Wi-Fi and 3G networks, and Android phones connect to all of the previous. This chapter explores the last pieces of Android hardware that tie the devices together and connect them with each other.

Overview of Android Hardware

Previous chapters have already explored the camera, sensors, and location hardware supported by Android. What remains is the hardware and software that enables an Android device to communicate with the outside world.

All Android devices have some capability to connect to the Internet; most have near-range wireless support for peripherals and many have the capability to place and receive calls over nationwide wireless networks. These are the core communication features common to Android:

- **Wi-Fi.** Wireless Internet access to local wireless access points and routers.

- **3G/4G.** Wireless Internet access to nationwide, digital data services.

- **LAN.** Wired Internet access via an Ethernet port and cable (Google TV only, for now).

- **Bluetooth.** Wireless serial and data access with other Bluetooth enabled devices, usually within a 30-to-75-foot range.

- **Voice Calls.** Calling to and from any phone number.

- **Text Messaging (SMS).** Short text message communication to and from phone numbers.

- **Multimedia Messaging (MMS).** Short messages with attachments to and from phone numbers.

Android provides APIs to interact with each of these service layers and features, which you can use to extend the functionalities of your app.

The Phone Itself

As you consider the additional features you want to include in your app, you frequently have to decide how much you want your app to do and what you are content to allow other apps and services to provide. In the case of calls and messaging, Android provides easy ways to pass requests to call or message contacts and numbers to native functions. If you want to control how Android handles these functions in more detail, it is possible to define your own procedures.

Suppose your app includes a contact list with phone numbers that you want to allow the user to dial directly. As you remember from Chapter 5, you can associate an onClick() event with a static Intent:

```
Intent i = new Intent( Intent.ACTION_CALL, Uri.parse( "tel:5551234567"
) );

startActivity( i );
```

The native dialer is normally responsible for listening for the call request and carrying the call through completion. Similarly, you can send a text message:

```
Intent i = new Intent( Intent.ACTION_SENDTO, Uri.parse(
    ➥"sms:5551234567" ) );
i.putExtra( "sms_body", "This message is for you" );
startActivity( i );
```

With a little extra syntax, you can send a multimedia message (MMS)—a text message with a media attachment:

```
Intent i = new Intent( Intent.ACTION_SEND, attached_Uri );
i.putExtra( "sms_body", "This image is for you");
i.putExtra( "address", "07912355432" );
i.putExtra( Intent.EXTRA_STREAM, Uri.parse( "content://media/
    ➥external/images/media/1" ) );
i.setType( "image/png" );
startActivity( i );
```

Working with Calls

If you need more control over these actions, you can intercept the native apps with your own code. You must first declare one or more of these permissions in AndroidManifest.xml to use the necessary features:

- **CALL_PHONE.** Grants permission to bypass native dialer and place outbound calls.

- **CALL_PRIVILEGED.** Grants permission to make outgoing and emergency calls, like 911.

- **PROCESS_OUTGOING_CALLS.** Grants permission to modify, monitor, or abort outgoing calls.

- **READ_PHONE_STATE.** Grants read-only permission to the type, number, and state of the phone.

Additionally, you must instruct the manifest to those activities that will implement the features by declaring intent filters.

If we wanted to do an Activity named Dialer, we could do the following:

```
<activity
android:name=".Dialer"
```

continues

```
android:label="@string/app_name">
    <intent-filter>
            <action android:name="android.intent.action.CALL_
BUTTON" />
            <category android:name="android.intent.category.
DEFAULT" />
    </intent-filter>
    <intent-filter>
            <action android:name="android.intent.action.VIEW" />
            <action android:name="android.intent.action.DIAL" />
            <category android:name="android.intent.category.
DEFAULT" />
            <category android:name="android.intent.category.
BROWSABLE" />
            <data android:scheme="tel" />
    </intent-filter>
</activity>
```

For most purposes, the only actions you need to reference are:

- **ACTION_CALL.** Performs a call based on the provided data. Cannot be used for emergency numbers.

- **ACTION_CALL_BUTTON.** Performs a call in response to a defined "call" button.

- **ACTION_DIAL.** Prepares a call based on the provided data. Displays a screen with the number to dial, which allows the user to explicitly begin the call.

- **ACTION_NEW_OUTGOING_CALL.** Broadcast action to notify BroadcastReceivers that an outgoing call is about to happen.

- **ACTION_VIEW.** Opens some content for viewing, like a contact or phone number.

Implementing these Intents, you can call these actions and add custom and extra data. In most cases, it is sufficient to allow Android to manage the call stack—that is, to decide for itself the best way to execute the call action. However, it is possible to replace the call stack with your own—useful if you need to develop your own phone app.

There are other reasons to integrate with the phone service layer.

Let's say we want to create an app that tells users three things:

- How long they spent on the phone
- With whom they spoke
- How frequently they spoke

To do so, we need to access the phone's state and interact with the TelephonyManager service:

```
String myTelephony = Context.TELEPHONY_SERVICE;
TelephonyManager _TelephonyManager = (TelephonyManager)
  ➥getSystemService( myTelephony );
```

The TelephonyManager class includes a number of methods that you can call to retrieve information about the device and its state, as shown in the following table.

Common TelephonyManager Methods

Method	Description	Return Examples
getCallState()	Returns the state of a call	CALL_STATE_RINGING, CALL_STATE_IDLE
getDataActivity()	Returns current data activity	DATA_ACTIVITY_INOUT, DATA_ACTIVITY_NONE
getDataConnected()	Returns the data connection state	DATA_CONNECTING, DATA_CONNECTED
getLine1Number()	Returns the phone number of the device	A phone number
getNetworkOperator()	Returns a numeric code for the carrier	A carrier number code
getNetworkType()	Returns the type of carrier network	NETWORK_TYPE_CDMA, NETWORK_TYPE_GPRS, NETWORK_TYPE_HSDPA
getPhoneType()	Returns the type of phone	PHONE_TYPE_CDMA, PHONE_TYPE_GSM, PHONE_TYPE_NONE
getVoicemailNumber()	Returns the registered voicemail number as a string	A registered voicemail number

You can then implement some simple logic to determine whether or not to place a call:

```
int callState = telephonyManager.getCallState();
switch ( callState ) {
      case ( TelephonyManager.CALL_STATE_RINGING ):
      //Do nothing
      break;
      case ( TelephonyManager.CALL_STATE_IDLE ):
      //Good to go, make the call
      break;
      case ( TelephonyManager.CALL_STATE_OFFHOOK ):
      //Toast the user
      break;
      default:
      break;
}
```

The last piece of working with Android as a phone is to manage or react to changes in the call state. With access to all of the information about the device's connection, you can implement a "Can You Hear Me Now?" app that prompts the user to ask that question with every negative change in network connections. You can also listen for events, such as incoming calls, and write your own reactions to them.

This type of access is enabled by the PhoneStateListener class. This class contains a collection of callback methods necessary to react to changes in the state of the phone. As with the TelephonyManager class, their purpose is easily derived from their names:

- **onCallStateChanged().** When device call state changes.

- **onCellLocationChanged().** When device cell location changes.

- **onDataActivity().** When data activity state changes.

- **onDataConnectionStateChanged().** When connection state changes.

- **onMessageWaitingIndicatorChanged().** When the message-waiting indicator changes.

- **onServiceStateChanged().** When device service state changes.

- **onSignalStrengthsChanged().** When network signal strengths changes.

To implement a listener to react to changes in phone state, you can declare:

```
PhoneStateListener _PhoneStateListener = new PhoneStateListener() {
    public void onCallStateChanged( int state, String
➥incomingNumber ) {}
    public void onServiceStateChanged( ServiceState serviceState )
{}
    public void onSignalStrengthsChanged( int asu ) {}
};
TelephonyManager.listen( phoneStateListener,  PhoneStateListener.
➥LISTEN_CALL_STATE |
PhoneStateListener.LISTEN_SERVICE_STATE |
PhoneStateListener.LISTEN_SIGNAL_STRENGTHS );
```

As always, anticipate changes to the Activity life cycle and implement onPause() and onResume() events to manage your app's use of system resources. To suspend your listening activity onPause(), use:

```
TelephonyManager.listen( PhoneStateListener, PhoneStateListener.
➥LISTEN_NONE);
```

With these methods and callbacks, you can implement any number of app-specific behaviors. Whether reacting to changes in signal quality or providing custom dialing features, the TelephonyManager and PhoneStateListener classes give you everything necessary to get started.

Send SMS and MMS Messages

Permissions for SMS communication are simpler and require only SEND_SMS:

```
<uses-permission android:name="android.permission.SEND_SMS"/>
```

Just as with the phone services, you can interact directly with text services by instancing the SmsManager class. The class is much simpler than the TelephonyManager class, and offers a few, quick methods to work with SMS:

- **divideMessage().** Splits a message into several smaller messages, each smaller than the maximum SMS message size.

- **getDefault().** Returns the default instance of the SmsManager.

- **sendDataMessage().** Send a data based SMS to a specific application port.

- **sendMultipartTextMessage().** Send a multipart text message.

- **sendTextMessage().** Send a text message.

With these methods, you can now send a simple SMS communication:

```
SmsManager _SmsManager = SmsManager.getDefault();
String contact = "5551234567";
String message = "Here is an SMS message from code";
SmsManager.sendTextMessage( contact, null, message, null, null
➥);
```

Just as with calls, you can listen for incoming SMS messages. SMS does not use the same service stack as the phone, which means that a dedicated service with predefined methods is not available to do this work. Instead, you need to expand or create a BroadcastReceiver class to listen for the relevant broadcast events:

```
private SmsManager _SmsManager;
public void onReceive( Context context, Intent i ) {
        if( i.getAction().equals( "android.provider.Telephony.SMS_
➥RECEIVED" ) ) {
                SmsManager = SmsManager.getDefault();
                Bundle bundle = i.getExtras();
                    if( bundle != null ) {
                            //There may be more than one, get all of
➥them
                            Object[] pdu = (Object[]) bundle.get(
➥"pdus" );
                            SmsMessage[] sms = new SmsMessage[pdu.
➥length];

                            for( int i = 0; i < pdu.length; i++ )
                            sms[i] = SmsMessage.createFromPdu(
➥(byte[]) pdu[i] );

                            for( SmsMessage txt: sms ) {
                                String msg = sms.getMessageBody();
                                String sender = sms.
➥getOriginatingAddress();

                                if( msg.toLowerCase().startsWith(
➥"shopping" ) ) {

                                        sms.sendTextMessage(sender,
➥null, "I'm on my way!", null, null);
                                }
                        }
                }
        }
}
```

This sample code would listen for incoming SMS messages that begin with "shopping" and then immediately send a reply to the sender, "I'm on my way!"

The emulator includes tools to simulate incoming calls and text messages. With the emulator running, launch the DDMS perspective or open the Emulator Control view, as shown in Figure 14.1.

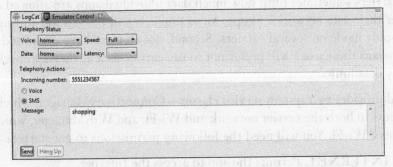

Figure 14.1: *The Emulator Control view is used to simulate calls.*

Network Interfaces

In the next few years, voice and data services will merge, and you will interact with the phone's core services as pure data. Some of this data will contain voice communication, some short messages and e-mail, and other browsing and streaming media. Until then, devices are still separated into two distinct departments: the voice department and the data department.

With voice services behind you, the next bundle of services handle the communication of data. From the data signal to a carrier, to the Wi-Fi signal with a local hotspot, to peer-to-peer data connections over Bluetooth, the average Android device is packed with antennas and receivers.

Most current Android devices support three versions of Wi-Fi: B, G, and N. Bluetooth support ranges from versions 2.0, 2.1, and 3.0—each with different hardware specifications. Android bundles each of these services into discreet classes, which eliminates the need for you to worry about how the service is implemented, allowing you the freedom to plug and play.

Wi-Fi vs. 3G

In many cases, you won't actually care how the device connects to the Internet. You simply want to be able to test for connectivity and get about the rest of your work. There are cases, however, where it makes sense to distinguish between the two primary ways a device can get "online."

First, carriers sometimes limit how much data individual apps are allowed to transmit over their data networks. Skype, for example, is still limited to running in Wi-Fi–only mode on several carriers. Second, not all users have unlimited data plans, which means these users will prefer not to use carrier data as they approach their monthly data limits.

Android provides two system service classes—ConnectivityManager, which manages data access to both the carrier network and Wi-Fi, and WifiManager, which manages the state of Wi-Fi. You will need the following permissions to get started:

- **INTERNET.** Permits the app to access the Internet.
- **ACCESS_NETWORK_STATE.** Permits read access to the connection state.
- **CHANGE_NETWORK_STATE.** Permits modify access to the connection state.
- **ACCESS_WIFI_STATE.** Permits read access to the Wi-Fi state.
- **CHANGE_WIFI_STATE.** Permits modify access to the Wi-Fi state.

The ConnectivityManager is instanced from the system service:

```
ConnectivityManager _ConnectivityManager =
(ConnectivityManager) getSystemService( Context.CONNECTIVITY_SERVICE
);
```

Inside the ConnectivityManager class, the following methods are available:

- **getActiveNetworkInfo().** Returns information about active network interfaces.
- **getAllNetworkInfo().** Returns information about all network interfaces.
- **getBackgroundDataSetting().** Returns true if background data usage is enabled; false if otherwise.
- **getNetworkInfo().** Returns network info by type (Mobile, Wi-Fi, WiMAX).

- **isNetworkTypeValid().** Returns true if the network type is valid.

- **requestRouteToHost().** Returns true if a network route exists to a specific address using a given network type.

- **startUsingNetworkFeature().** Instructs the networking system to begin using the named feature.

- **stopUsingNetworkFeature().** Instructs the networking system to stop using the named feature.

With getBackgroundDataSetting(), you can evaluate whether the user permits apps to continue using data services in the background. It is up to your app to respect this setting, as Android does not automatically enforce it for you.

Changes in the value of this setting send a broadcast. Here's how you find out if the setting has changed:

```
registerReceiver( new BroadcastReceiver() {
        @Override
        public void onReceive( Context context, Intent intent ) {
            // React to change in preferences
        },
        new IntentFilter( ConnectivityManager.ACTION_BACKGROUND_DATA_
    SETTING_CHANGED ) );
}
```

If you have ever used the Pandora app to listen to music on the go, you have probably noticed changes in audio quality as you move through areas with weak network connections. This is possible by listening for changes in the network state. Using the same structure as above, register an IntentFilter to listen for ConnectivityManager. CONECTIVITY_ACTION. This will respond to broadcast changes, which include:

- **EXTRA_EXTRA_INFO.** Contains a lookup key for optionally supplied extra information about the network state.

- **EXTRA_IS_FAILOVER.** True if the ConnectivityManager has failed connection on one network and is attempting to roll over to another network.

- **EXTRA_NETWORK_INFO.** Contains a lookup key for a NetworkInfo object.

- **EXTRA_NO_CONNECTIVITY.** Indicates whether no network connectivity is possible.

- **EXTRA_OTHER_NETWORK_INFO.** The lookup key for a NetworkInfo object containing additional, optional information.

- **EXTRA_REASON.** Provides a reason why an attempt to connect to a network failed.

If you receive notification that a failover has occurred, you can query the ConnectivityManager to see which network type is active. Use the following code to do this:

```
NetworkInfo _NetworkInfo = _ConnectivityManager.
  ➥getActiveNetworkInfo();
int networkType = networkInfo.getType();
switch( networkType ) {
        case ( ConnectivityManager.TYPE_MOBILE ):
        // React to lower possible speeds
        break;
        case ( ConnectivityManager.TYPE_WIFI ):
        // React to higher bandwidth available
        break;
        default:
        break;
}
```

If you just need to assess the state of a particular network service, active or not, you can call getNetworkInfo() on the service type:

```
NetworkInfo mobileNetwork = _ConnectivityManager.getNetworkInfo(
  ➥ConnectivityManager.TYPE_MOBILE );
```

Your users will frequently have simultaneous access to more than one network type, in which case you can explicitly prefer one over another.

```
_ConnectivityManager.setNetworkPreference( NetworkPreference.PREFER_
  ➥WIFI );
```

If necessary, you can enable and disable network types by calling setRadio(), though you should request the user's permission through an alert before doing so.

```
_ConnectivityManager.setRadio( NetworkType.WIFI, true );
```

Wi-Fi

While the ConnectivityManager provides blanket access to all network types, the WifiManager has a more specific focus, namely joining Wi-Fi networks. Unlike communicating with a carrier, where an Android device is invisible authenticated, most Wi-Fi hotspots originate from a single transceiver, each with different names and security measures.

Android provides built-in utilities to manage a device's connection to these hotspots, but it is sometimes desirable to manage this process manually in your app—or even to design a new Wi-Fi management app from scratch.

Begin by instancing the WifiManager system service:

```
WifiManager _WifiManger = (WifiManager) getSystemService( Context.WIFI_
➥SERVICE );
```

You can register an IntentFilter to listen for broadcast changes in the Wi-Fi state registered by the WifiManager. These include:

- **WIFI_CHANGED_STATE_ACTION.** Indicates that the hardware state is changing between enabled, enabling, disabled, disabling, or unknown. You can react to this by attempting to re-enable the Wi-Fi:

```
if( !_WifiManager.isWifiEnabled() )
if( _WifiManager.getWifiState() != WifiManager.WIFI_STATE_ENABLING )
_WifiManager.setWifiEnabled( true );
```

- **SUPPLICANT_CONNECTION_CHANGE_ACTION.** Indicates that the connection with a hotspot has changed. Either a new access point has been joined or the previous connection has been lost.

- **RSSI_CHANGED_ACTION.** Indicates a change in the signal quality to an access point.

The WifiManager class provides even more functionality, allowing you to scan available access points, evaluate signal strength and quality, and join networks. For most purposes, you will not need to perform any of these actions. However, if you do want to build your own Wi-Fi management service, the Android SDK provides all of the necessary tools to do so.

Working with Bluetooth

Perhaps the most under-utilized hardware feature available on your Android device is the *Bluetooth* receiver. Most commonly, Bluetooth headsets enable wireless audio connections with the device. Bluetooth's potential is actually much greater. While it is most widely used with peripherals, such as mice, keyboards, and controllers, it can be used to communicate any type of information.

You can think of the Bluetooth connections created by an Android device as network connections. These networks typically only include two members: the Android device and something else, but this is really a network connection, treated much the same as any other wireless network.

DEFINITION

Bluetooth is a wireless communication standard that allows data communication with any other Bluetooth-enabled device within range. Range is typically between 30 and 75 feet.

As always, you will need permission to work with Bluetooth on Android. Two permissions exist:

- **BLUETOOTH.** Grants permission to use/perform any Bluetooth communication.

- **BLUETOOTH_ADMIN.** Grants permission to discover devices and enable/disable the service. The advanced admin functions should not be used unless you are designing a Bluetooth management app.

Android has implemented Bluetooth through a series of classes, each of which you will need to implement.

- **BluetoothClass.** Provides a read-only set of properties that describe the generic capabilities of the device. This is not specific to Bluetooth hardware installed, but it useful as a guideline.

- **BluetoothAdapter.** Provides access to the Bluetooth radio. Allows discovery of other Bluetooth devices, querying paired devices, instancing a BluetoothDevice, and creating a BluetoothServerSocket to communicate with other devices.

- **BluetoothDevice.** Represents a remote Bluetooth device. Uses **BluetoothSocket** to connect with a remote device or to find out information about the remote device.

- **BluetoothSocket.** Allows communication between Bluetooth devices via InputStream and OutputStream.

- **BluetoothServerSocket.** Provides the server socket to listen for incoming requests. When two devices are connected, one must implement a server socket.

Your starting point for working with Bluetooth is the BluetoothAdapter, which you need to instance to gain access to the radio. Check first to ensure that the BluetoothAdapter instance is not null. If it is, the device does not support Bluetooth.

```
BluetoothAdapter _BluetoothAdapter = BluetoothAdapter.
➥getDefaultAdapter();
if( null == _BluetoothAdapter )
// Bluetooth is not supported
```

Next, you need to ensure the Bluetooth radio is enabled. Request to enable it, and the user will see a pop-up window asking for permission to enable the radio.

```
if( !_mBluetoothAdapter.isEnabled() ) {
        Intent i = new Intent( BluetoothAdapter.ACTION_REQUEST_ENABLE
➥);
        startActivityForResult( i, REQUEST_ENABLE_BT );
}
```

The Bluetooth radio is also enabled when you request that the Android device be discoverable to other devices. If this is something you plan to do regularly, you can skip enabling the radio and request for discoverability mode instead. For power conservation, devices remain discoverable for a short period of time. Three hundred seconds, or five minutes, is the maximum permitted.

```
Intent i = new Intent( BluetoothAdapter.ACTION_REQUEST_DISCOVERABLE
    );
i.putExtra( BluetoothAdapter.EXTRA_DISCOVERABLE_DURATION, 300 );
startActivity( i );
```

With the Bluetooth radio enabled, you next want to do one of two things: connect to a *paired* device or find a new device. To query paired devices, call getBondedDevices(). This returns a BluetoothDevice Set, which you can iterate:

```
Set<BluetoothDevice> allDevices = _BluetoothAdapter.
➥getBondedDevices();
if( pairedDevices.size() > 0 ) {
        for( BluetoothDevice pair : allDevices ) {
                // Evaluate the device or take an action using pair.
➥getAddress()
        }
}
```

DEFINITION

Paired means a Bluetooth device has been discovered by and connected to another Bluetooth-capable device. Once paired, the basic device information (name, address) is saved and communication with the device is possible.

The only piece of information you need to establish a connection to a Bluetooth device is its MAC address—the value returned from the getAddress() method. If getBondedDevices() does not return a desired device, or if you know you need to find a new Bluetooth device, you must try to discover devices in the area.

PITFALL

Bluetooth device discovery is an energy-expensive operation. It should not be done if already connected to a device, as this will degrade the connection. Once you have found a target MAC address, cancel the discovery operation. Be sure discovery is cancelled onPause() as well.

Initiating device discovery is as simple as calling the startDiscovery() method. This process polls for devices, running at approximately 10-second intervals. Discovered devices broadcast the ACTION_FOUND intent, so you must register a BroadcastReceiver to be notified. Use the following code to do this:

```
private final BroadcastReceiver _BroadcastReceiver = new
    BroadcastReceiver() {
        public void onReceive( Context context, Intent intent ) {
                String found = intent.getAction();
                if ( BluetoothDevice.ACTION_FOUND.equals( found ) ) {
                        BluetoothDevice device = intent.
    getParcelableExtra(BluetoothDevice.EXTRA_DEVICE);
                        // Do something with found device(s) device.
    getAddress()
                }
        }
};
IntentFilter filter = new IntentFilter( BluetoothDevice.ACTION_FOUND
    );
registerReceiver( _BroadcastReceiver, filter );
```

When two Bluetooth devices connect, one must connect as a server and the other must connect as a client. The server connection involves three steps:

1. Obtain a BluetoothServerSocket by calling the listenUsingRfcommWithServiceRecord() method. This method takes the name of your service and a *UUID*.

2. Begin listening for connection requests by calling the accept() method. Once complete, accept() returns a connected BluetoothSocket.

3. Close the server socket by calling the close() method. This does not close BluetoothSocket.

DEFINITION

A **Universally Unique Identifier (UUID)** is a random string that uniquely identifies your application's Bluetooth service.

This process should not occur in your main Activity thread, as many of the Bluetooth method calls are blocking calls—that is, no other action in the same thread can occur until these methods complete. Threads are discussed in greater detail in Chapter 16.

Your server connection thread should now look like this:

```
private class OpenBluetoothThread extends Thread {
        private final BluetoothServerSocket _BluetoothServerSocket;
        public OpenBluetoothThread () {
                BluetoothServerSocket tmpSocket = null;
                try {
                        tmpSocket = _BluetoothAdapter.
        listenUsingRfcommWithServiceRecord( "My Service", some_UUID );
                }
                catch (IOException e) { }
                BluetoothServerSocket = tmpSocket;
        }
        public void run() {
                BluetoothSocket btSocket = null;                while
        (true) {
                        try {
                                btSocket = _BluetoothServerSocket.
        accept();
                        }
                        catch (IOException e) {
                                break;
                        }
                        if (btSocket != null) {
                                // Take some action with the socket
```

continues

```
                        BluetoothServerSocket.close();
                        break;
                }
        }
}
public void cancel() {
        try {
                BluetoothServerSocket.close();
        }
        catch( IOException e ) { }
        }
}
```

Connecting as a client to a server device follows a similar path.

1. Using BluetoothDevice, instance a BluetoothSocket by calling createRfcommSocketToServiceRecord(). This method takes a UUID which must match the server socket. UUIDs can be stored as resource string values to ensure matching.

2. Start the connection by calling the connect() method.

Once you have established the communication channel between the devices, you are free to exchange any kind of data. The Android Developer's Guide provides far greater detail on how to move files between devices, exchange content, or share libraries.

Testing Bluetooth communication requires more than one physical device, as you will need to verify both the server and the client side of the connection. The emulator will not allow you to accurately test Bluetooth functionality.

Whether you need to know the state of your devices connection to the Internet or other devices, Android provides services to feed meaningful state information into your app. You can use this information to optimize the app experience for your users, or you can enhance your app's communication ability by extending Bluetooth and Wi-Fi services.

The Least You Need to Know

- Using Action Intents is the simplest way to place a call or send an SMS.
- The **TelephonyManager** and **SmsManager** classes provide access to the system services that manage calls and messaging.
- The **ConnectivityManager** and **WifiManager** provide access to the system services that manage network connectivity.
- Bluetooth interaction requires an instance of **BluetoothAdapter,** a **BluetoothServerSocket,** and a **BluetoothSocket.**
- Most system services will broadcast changes in their status, which can be monitored with **BroadcastReceivers.**

The Least You Need to Know

Using Action Intents is the simplest way to place a call or send an SMS.

The TelephonyManager and SmsManager classes provide access to the system services that manage calls and messaging.

The ConnectivityManager and WifiManager provide access to the system services that manage network connectivity.

Bluetooth interaction requires an instance of BluetoothAdapter, a BluetoothServerSocket, and a BluetoothSocket.

Most system services will broadcast changes in their status, which can be monitored with BroadcastReceivers.

Increasing Your Application Scope

The sky is the limit when it comes to apps, and the Android platform can make an app a unique experience for each user. Part 4 discusses using other elements to make your app more complex, including accessing databases using SQL and utilizing multitasking.

A Touch of Locale

In This Chapter

- Customizing your app by region and language
- Translating content on the fly using Google's translation servers
- Preparing for critical changes in the Activity life cycle
- Working with multi-touch

Only in the last decade have users been able to buy cheap, portable devices that will work in almost any country in the world. Now, not only can you buy mobile Android phones that will operate worldwide, but you can write apps that will run on any Android phone, anywhere.

In this chapter, you learn how to localize your app by language and region. Additionally, you learn how to cope with smaller changes in orientation, like switching between landscape and portrait mode. Finally, you sample some of the possibilities of multi-touch in Android.

Localization

As the song goes, "it's a small world after all." With few exceptions, by publishing your app to the web, you have made it available to anyone from any country to download and enjoy. To provide the best possible experience to all your potential users, it is sometimes useful to customize the language and other resources your app uses to reach the broadest possible audience.

Localization is the process of providing additional resources for your app so that it matches the language and region of its user. When users set up their Android devices, language selection is one of the first steps. Android automatically detects country and region from the network.

As you learned in Chapter 3, you can provide alternate resources used to customize your app by orientation and time of day. The same is true for languages. By providing additional resources, your app can automatically adjust to the language and region of your users.

Managing Resources

Specifying language and region resources follows the same logic as providing optimized resources for low- to high-resolution screens. For example, icon.png might be your app's home screen icon. The system default image should live in /res/drawable/icon.png and you might specify optimized versions at /res/drawable-ldpi/icon.png and /res/drawable-hdpi/icon.png.

In all cases, the resource name icon.png stays the same. The contents of the resource differ by location. Android selects the most appropriate version to use at runtime.

The same concept applies to locales that might have different string value content for languages and regions, and that might use different image drawables for countries. Due to limitations in space, the Android SDK does not immediately support all possible locales.

Here are the locales currently supported by the Android SDK. They are arranged by language, location, and resource name.

Locales Supported by the Android SDK

Language	Locale	Resource Name
Chinese	PRC	zh_CN
Chinese	Taiwan	zh_TW
Czech	Czech Republic	cs_CZ
Dutch	Netherlands	nl_NL
Dutch	Belgium	nl_BE
English	US	en_US
English	Britain	en_GB
English	Canada	en_CA
English	Australia	en_AU

Language	Locale	Resource Name
English	New Zealand	en_NZ
English	Singapore	en_SG
French	France	fr_FR
French	Belgium	fr_BE
French	Canada	fr_CA
French	Switzerland	fr_CH
German	Germany	de_DE
German	Austria	de_AT
German	Switzerland	de_CH
German	Liechtenstein	de_LI
Italian	Italy	it_IT
Italian	Switzerland	it_CH
Japanese	Japan	ja_JP
Korean	Korea	ko_KR
Polish	Poland	pl_PL
Russian	Russia	ru_RU
Spanish	Spain	es_ES

GOOGLE IT

Android supports many more languages than these, but the SDK does not implement every variation. Refer to the Android Open Source project for a complete reference at http://source.android.com.

At the simplest level, you may want to provide alternate translations for the core text of your app. Start with the application name. Assuming that English is the default language of your app, your /res/values/strings.xml resource might define app_name="Simply Recipes". You can then define the app name in Spanish and French.

```
In /res/values-es/strings.xml:
        <string name="app_name">Simplemente Recetas</string>
In /res/values-fr/strings.xml:
        <string name="app_name">Simplement Recettes</string>
```

If a user has selected French or Spanish as a default language, he will see the corresponding value as the name of the app. All other values default to those defined in the default values resource folder /res/values.

PITFALL

While you can specify a language without a locale, as in /res/values-fr/string.xml, you cannot specify a locale without a language. A resource named /res/values-FR/strings.xml would generate a compilation error. Locales require a language identifier; languages do not require a locale.

If you do not provide the default resources, it is easy to generate fatal errors for your users. You could define the app's name only in /res/values-en. Your app would work perfectly for all English-speaking users, but as soon as a German user attempted to load your app, they would receive a fatal error. It is best to first define all possible resources in the default locations and then build out support for other locales.

Alternate Resources

The default, or root, resource qualifiers are those with the least specificity in their names. For drawables, this will be /res/drawable, as drawable has the least possible definition. For values, this will be /res/values. Android does not assume that the resources in these locations will be for U.S. English—these could contain content for any locale or language.

You probably do not need to support more than English or your own native language for the first version of your app. As you start collecting feedback from users, you will have the data needed to design a localization strategy. You can use the data to make the following decisions for your app:

1. Which languages and locales does your application need to support? Only support these groups until you have time and resources and user demand to support more.

2. Which language and locale will be your default? If U.S. English is your default country and language, everything in your default resources folders should target these.

3. What content will your app display if you have not provided localization for a language or locale? Will you evaluate a user's language and region in order to display something other than the default resources?

4. Which resources actually need to be localized? Certain types of values like currency, dates and times, and weights and measures make sense to prepare for localization. Neutral images and sounds, text which is not displayed in the user interface, colors, and styles may not need different versions.

5. Are there any aspects that may be offensive to certain cultures? If you hope to reach users across the world, either design your app to appeal to most cultures or design separate themes and styles to target specific regions.

6. Have you put your data in remote places? In both Java and XML, use external resources as much as possible. Prefer using the R class to retrieve strings, drawables, and layouts instead of hard-coding resource values.

7. How accurate are your translations? If you do not speak all of the languages needed to localize your app, you must get accurate translations from someone who does and format them correctly.

Once you have completed the process of preparing localized resources, you can begin to test and debug your app in the emulator. See Figure 15.1.

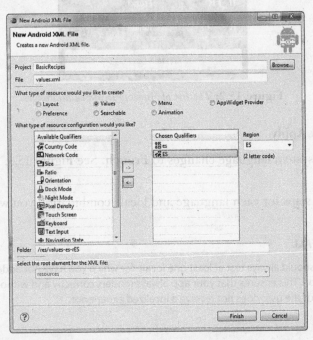

Figure 15.1: *Eclipse automatically creates language and region specific resource files for new Android XML files.*

1. Start the emulator.
2. Click the **Applications** icon.
3. Select **Custom Locale.**
4. Click **French.** See Figure 15.2.

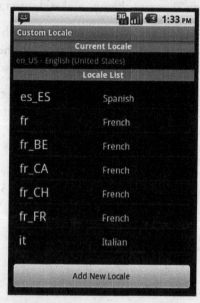

Figure 15.2: *The list of Android localized languages.*

5. Click **Apply.**
6. The system language changes to **French.** See Figure 15.3.

Repeat these steps for each language and locale combination you wish to test.

PITFALL

You should always test at least one locale for which you have not added localization. This ensures that your app always renders correctly and without error, even if the user does not receive a localized experience.

Figure 15.3: *The Android localized in French.*

Performing Translation in Code

Supporting even two or three locales can become quite time consuming, as you can imagine. As your app grows, the required maintenance for localization expands and each resource for each locale has to be updated and tested. Your app can easily become quite large with locale specific changes.

Some of this is unavoidable. If you have region-specific images like flags or national monuments, a localized app needs to have alternate drawables for different regions. Text translations, however, have come a long way. Google offers translation services through browser toolbars, a dedicated site, and the Google Translate Android app.

GOOGLE IT

If you have not used Google Translate before, you should find http://translate. google.com and enter a selection of foreign newspapers; the German Zeit Online for example: http://www.zeit.de. While not perfect, Google Translate is accurate enough to convey much of the meaning of the original.

While an official Android API does not yet exist, you can make use of a widely supported, unofficial API to begin translating text automatically through Google's servers. To proceed, follow these steps:

1. Make sure your app has **INTERNET** permission declared in AndroidManifest.xml.

2. Visit http://code.google.com/p/google-api-translate-java.

3. Download the latest version of the google-api-translate-java JAR file—currently **google-api-translate-java-0.92.jar.**

4. Store this in an easy to reference location, such as /<Your Eclipse Workspace>/shared.

5. In Eclipse, select your Project and hit **ALT + Enter** to open **Properties.**

6. Select **Java Build Path** and click the **Libraries** tab. See Figure 15.4.

7. Click **Add External JARs.** See Figure 15.4.

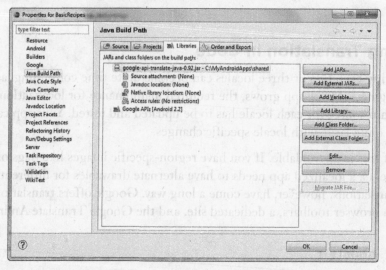

Figure 15.4: *The Java Build Path.*

8. Select the location of the translate API JAR and click **OK.**

9. The new APIs are now visible in Package Explorer under Referenced Libraries, as shown in Figure 15.5.

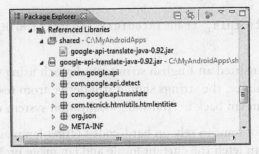

Figure 15.5: *The Referenced Libraries in the Package Explorer.*

Now you are ready to translate some text in Java. First, import the references to the new APIs:

```
import com.google.api.translate.Language;
import com.google.api.translate.Translate;
```

Define the new Activity class:

```
import android.app.Activity;
import android.os.Bundle;
import android.widget.Toast;
public class TranslateMe extends Activity {
        @Override
        public void onCreate(Bundle savedInstanceState) {
                super.onCreate(savedInstanceState);
                setContentView(R.layout.main);
                doTranslate();
        }
}
```

Implement the doTranslate() method that displays a toast dialog with the translation of your text:

```
private void doTranslate(){
        String translatedText = null;
        try {
                Translate.setHttpReferrer( "http://www.mysite.com" );
                Translate.validateReferrer();
                translatedText = Translate.execute( "Welcome to this
app!", Language.ENGLISH, Language.FRENCH );
        }
        catch( Exception e ) {
                e.printStackTrace();
                translatedText = "An error occurred during
translation";
```

continues

```
        }
    Toast.makeText( this, translatedText, Toast.LENGTH_LONG ).show();
    }
```

Voilà! You have translated an English string into French using Google's translation servers. As best practice, the strings should still derive from resource values. This ensures that you can fail back to Android's localization system if necessary.

Likewise, you do not want to rely on hard-coded target languages. Using the Android Locale class, you can fetch the current locale and language preferences of the user. You can then define a current language string as:

```
String currentLanguage = Locale.getDefault().getDisplayLanguage().
    ➥toUpperCase();
```

Modify the Translate.execute() method to use currentLanguage:

```
Translate.execute( "Welcome to this app!", Language.ENGLISH,
    ➥Language.fromString( currentLanguage ) );
```

As long as the user has a responsive Internet connection, you can translate almost any text into any supported language—over 50 and counting.

Runtime Changes

It is difficult to imagine computers without mice, so much so that few users even think about using computers without trackpads or mice. Similar expectations about standard smartphone functionality and behavior are converging now. Perhaps the simplest expectation of a smartphone is that it automatically transitions between landscape and portrait mode, and that this transition should happen as fast as possible.

While Android handles transitions like this automatically, it does so by preempting the Activity life cycle—by calling onDestroy() and onCreate() in succession. Believe it or not, Android terminates and restarts your app every time a user changes the orientation. And not just for orientation changes. Any of the changes in the following table will instruct Android to take this action.

Configuration Changes That Trigger a Restart

Change	Usual Cause
locale	The user has changed the system language.
keyboard	The type of keyboard has changed, if the user has connected a Bluetooth or external keyboard.
keyboardHidden	The user has extended or retracted the hardware keyboard.
orientation	The user has rotated the device.
screenLayout	The user has changed the screen layout, perhaps by connecting another display.
fontScale	The user has changed the global font size.
uiMode	The user has plugged the device into a dock or day/night mode has changed.

For new developers, this behavior may be unexpected or even shocking, but it is really the most efficient way to ensure that the proper layout and resources are available to your app for the current orientation of the device. It does mean that you need to be aware of this behavior and plan your Activity life cycle accordingly.

For example, if your Activity begins by fetching a collection of the user's videos from a site like YouTube, this expensive operation would occur every time the user turned the device from side-to-side—hardly a desirable behavior. As a developer, you have two options:

- Store persistent data as an Object, which you save before onDestroy() and restore this data in onCreate() or onStart().

- Override Android's configuration change management and handle these changes within your app.

PITFALL

In most cases, it is more efficient and stable to allow Android to respond to configuration changes on its own. You should choose to override this only if truly necessary.

Efficiently Rotating the Screen

Users have come to expect speedy transitions. To guarantee this, you must reduce the overhead of your Activity's life cycle events by persisting the data retrieved from all expensive operations. Android provides two additional life cycle methods to assist you with this:

- **onRetainNonConfigurationInstance().** Android calls this method between onStop() and onDestroy(), and Android is smart enough to know when to make this call. You need only implement the method and pass an arbitrary object with your data, and Android will make sure this object is intact and available when you next need it. You should not use this object to store Views, Drawables, or anything tied to the Context or Activity as this could create a memory leak.

- **getLastNonConfigurationInstance().** This method is available in either onCreate() or onStart(), and can be used to restore your data from the persisted object. You must anticipate null returns here.

First, override the onRetainNonConfigurationInstance() method and return some object to be stored:

```
@Override
public Object onRetainNonConfigurationInstance() {
        return yourObject;
}
```

In this example, yourObject can be any visible property you have defined elsewhere in your code. Once set here, you are prepared to call getLastNonConfigurationInstance() and restore this data when your Activity is restarted:

```
private String yourObject = null;
@Override
public void onCreate(Bundle savedInstanceState) {
        super.onCreate(savedInstanceState);
        setContentView(R.layout.main);
        yourObject = (String)getLastNonConfigurationInstance();
        if( null != yourObject ) {
                //Restore the string to the proper context
        }
        else {
                //Your Activity is starting normally
        }
}
```

Remember that getLastNonConfigurationInstance() will return an object only if the Activity has restarted due to a configuration change. In all other cases, it will be null.

Manually Manage Configuration Changes

If your app does not need to change content for different configurations and if you have decided that Android's default behavior too greatly impacts your app's performance, then you can opt to manage these changes on your own. This involves two steps.

First, you must specify each Activity that will manage its own configuration changes in AndroidManifest.xml by defining the android:configChanges attribute:

```
<activity
android:name=".TranslateMe"
android:configChanges="orientation|uiMode|locale"
android:label="@string/app_name"
>
```

This instructs Android to call onConfigChanged() instead of initiating an Activity restart for orientation and uiMode changes only.

Second, you must implement the onConfigChanged() method in your Activity. Android will pass this method a Configuration object with the specific configuration that has changed.

```
@Override
public void onConfigurationChanged(Configuration configuration) {
        super.onConfigurationChanged(configuration);
        // Toast if the screen orientation has changed
        if(configuration.orientation == Configuration.ORIENTATION_
➥LANDSCAPE )
        {
                Toast.makeText( this, "Device has entered Landscape
➥orientation", Toast.LENGTH_SHORT ).show();
        }
        else if( configuration.orientation == Configuration.ORIENTATION_
➥PORTRAIT )
        {
                Toast.makeText( this, "Device has entered Portrait
➥orientation", Toast.LENGTH_SHORT ).show();
        }

        // Toast if the UI Mode has changed
```

continues

```
if( configuration.uiMode == Configuration.UI_MODE_TYPE_CAR )
{
        Toast.makeText( this, "Device has docked with a car
➥port", Toast.LENGTH_SHORT ).show();
}
else
{
        Toast.makeText( this, "Device has had an undefined UI
➥Mode change", Toast.LENGTH_SHORT ).show();
}

// Toast if the locale has changed
if( configuration.locale == Locale.ENGLISH )
{
        Toast.makeText( this, "Device has changed locale to
➥English", Toast.LENGTH_SHORT ).show();
}
else
{
        Toast.makeText(this, "Device has changed locale from
➥English", Toast.LENGTH_SHORT).show();
}
}
```

As you can see, writing the logic to handle all possible configuration changes can quickly become quite involved, and this sample does not cover more than a few possible Configuration values. For this reason, it is preferable to allow Android to handle as many of these changes as possible.

Multi-Touch

First popularized by the iPhone, gestures and multi-touch support have become standard and expected features of the smartphone. Android has supported multi-touch from its early stages and has incrementally added gestures support with each version. As of Android 2.2, gestures and multi-touch are on par with the iPhone and competing devices.

Gestures, such as pinch zoom, tap, swipe, and drag are rapidly evolving into a kind of smartphone sign language. Android handles each of these for you automatically, translating the user's input into method calls. In most cases, you do not need to interrupt this recognition; however, as you begin testing your app, you may find that pinch zoom does not always work the way you intend—or you may want to implement gestures of your own.

MotionEvents

Whether you realize it or not, you have already implemented multi-touch logic into your apps using the onTouch() method, which takes a View and a MotionEvent:

```
@Override
public boolean onTouch(View view, MotionEvent event) {
        // Your action
        return true;
}
```

The View class in Android implements two interfaces for responding to input or extending input functionality. The onTouch() method is a member of the onTouchListener() *event listener* interface. You can also use the onTouchEvent(), part of an *event handler* interface, which takes only a MotionEvent object.

> **DEFINITION**
>
> An **event listener** is a View interface which contains a single callback method, such as onClick() or onKey(). These methods target the View and not components of the View.
>
> **Event handlers** extend View components, such as Button or EditText widgets. Handlers provide method callbacks in the context of the parent widget.

If you wanted to define an Activity that consisted of a single ImageView with a drawable background, you might intercept the onTouch() event to modify the pinch zoom behavior. On the other hand, if you built a gallery of images in an Activity, you would more likely want to implement the onTouchEvent() to target actions for each View component.

Each of these methods takes a MotionEvent object. The MotionEvent class is your gateway to touch in the Android SDK. The most essential method of this class is getAction(), which returns the touch event. With the action, you can then call these methods to determine where the screen is touched and by how many fingers—also called pointers.

- **findPointerIndex().** Returns a number of the specified pointer index.
- **getPointerCount().** Returns a count of the pointers active on the screen.
- **getX().** Returns the x-axis coordinates of the first pointer or the specified pointer.

- **getXPrecision().** Returns the accuracy of the x-axis coordinate for a pointer.

- **getY().** Returns the y-axis coordinates of the first pointer or the specified pointer.

- **getYPrecision().** Returns the accuracy of the y-axis coordinate for a pointer.

Using the onTouchEvent() method and some relevant MotionEvent methods, you can build a simple drag and drop handler:

```
private float lastTouchedX;
private float lastTouchedY;
private float relocateX;
private float relocateY;
@Override
public boolean onTouchEvent( MotionEvent event ) {
        final int action = event.getAction();
        switch( action ) {
                case MotionEvent.ACTION_DOWN: {
                        final float x = event.getX();
                        final float y = event.getY();

                        // Store last position
                        lastTouchedX = x;
                        lastTouchedY = y;
                        break;
                }

                case MotionEvent.ACTION_MOVE: {
                        final float x = event.getX();
                        final float y = event.getY();

                        // Get the move distance
                        final float movedX = x - lastTouchedX;
                        final float movedY = y - lastTouchedY;

                        // Set move coordinates
                        relocateX += movedX;
                        relocateY += movedY;

                        // Store last position
                        lastTouchedX = x;
                        lastTouchedY = y;

                        break;
```

```
        }
    }
    return true;
}
```

With the relocateX and relocateY coordinates you have calculated, you can define a location on the screen to position a resource such as an image. This example only considers a single pointer or finger touching the screen at a time. In order to accept multi-touch input, you need to account for other MotionEvent actions:

- **ACTION_CANCEL.** The current gesture has stopped prematurely.

- **ACTION_DOWN.** The user has started pressing the screen. This event includes the starting location.

- **ACTION_UP.** The user has stopped pressing the screen. This includes the ending location.

- **ACTION_MOVE.** The user has changed the gesture action. This occurs between ACTION_DOWN and ACTION_UP.

- **ACTION_POINTER_DOWN.** The user has touched the screen with another finger(s).

- **ACTION_POINTER_UP.** The user has released the other finger(s) from the screen.

While most of the magic of multi-touch happens courtesy of Android, it is relatively simple to implement your own custom behavior.

Build Your Own Gestures

Of course, the number of predefined gestures in Android is fairly limited. To encourage developers and users to embrace gesture diversity, beginning with OS 1.6, the Android SDK includes a gesture builder. Here you can define your own custom gestures, which can be included in your app.

PITFALL

While all Android phones and tablets support multi-touch, the Android emulator does not. Any gestures created with the emulator should be well-tested on physical devices.

To begin, you must install the Gestures Builder API sample app and create your own gestures.

1. In Eclipse, create a new **Android Project.**
2. Select **Create project from existing sample.**
3. Select **Gestures Builder.**
4. Click **Finish** and start the project in the emulator.
5. Open the Apps menu and launch Gestures Builder.
6. Click **Add gesture.** See Figure 15.6.

Figure 15.6: *Adding gesture control to your app.*

7. Provide a name and a gesture by drawing a symbol with the mouse.
8. When finished, leave the emulator running and switch back to Eclipse.
9. Select **Window > Open Perspective > Other > DDMS.**
10. From the right-hand view, select **File Explorer.**
11. Expand **/mnt/sdcard** and select gestures.
12. Click the **Pull a file from the device** icon and place the gestures file into your project's raw resource folder, **/res/raw.**

Your custom gestures are now ready to use within your app. Implementing custom gestures involves far fewer steps.

1. Define a parent **GestureOverlayView** in your Activity's layout resource.

2. Implement the **onGesturePerformedListener** interface in your Activity's class.

3. Add the required **onGesturePerformed()** method.

The GestureOverlayView is not part of the standard widget package, so it must be declared explicitly. This view has three unique attributes worth mentioning:

- **gestureStrokeType.** Either single, for one finger, or multiple for multi-touch.

- **eventsInterceptionEnabled.** This instructs Android to begin listening for your custom gestures right away, which avoids confusion with standard Android OS gestures.

- **orientation.** This informs Android which scroll direction the child views will have, allowing Android to immediately recognize gestures in the opposite direction.

Now define a GestureOverlayView layout resource as:

```xml
<?xml version="1.0" encoding="utf-8"?>
<android.gesture.GestureOverlayView
        xmlns:android="http://schemas.android.com/apk/res/android"
        android:id="@+id/gesture_view"
        android:layout_width="fill_parent"
        android:layout_height="fill_parent"
        android:gestureStrokeType="multiple"
        android:eventsInterceptionEnabled="true"
        android:orientation="vertical"
>
<ListView
        android:id="@android:id/list"
        android:layout_width="fill_parent"
        android:layout_height="fill_parent"
/>
</android.gesture.GestureOverlayView>
```

Then prepare a simple Activity class to use the resource and handle some gestures. The raw gestures resource you created stores each gesture as a Prediction, which can be parsed as an ArrayList. Each Prediction has a score, which is the likeliness of a match for what the user has drawn and your custom gestures. Anything with a score of less than 1.0 is probably not a match.

You can now define a custom gesture-enabled Activity as:

```
import java.util.ArrayList;
import android.app.ListActivity;
import android.gesture.Gesture;
import android.gesture.GestureLibraries;
import android.gesture.GestureLibrary;
import android.gesture.GestureOverlayView;
import android.gesture.Prediction;
import android.gesture.GestureOverlayView.OnGesturePerformedListener;
import android.os.Bundle;
import android.widget.Toast;
public class GestureSample extends ListActivity implements
    OnGesturePerformedListener {
private GestureLibrary _GestureLibrary;
@Override
        public void onCreate( Bundle savedInstanceState ) {
                super.onCreate( savedInstanceState );
                setContentView( R.layout.main );

                GestureLibrary = GestureLibraries.fromRawResource(
    this, R.raw.gestures );
                if( !_GestureLibrary.load() )
                finish();

                GestureOverlayView gestureOV = (GestureOverlayView)
    findViewById( R.id.gesture_view );
                gestureOV.addOnGesturePerformedListener( this );
        }
        public void onGesturePerformed( GestureOverlayView overlay,
    Gesture gesture ) {
                ArrayList<Prediction> predictions = _GestureLibrary.
    recognize( gesture );
                if ( predictions.size() > 0 ) {
                        if (predictions.get( 0 ).score > 1.0)
                        Toast.makeText( this, "You performed the " +
    predictions.get( 0 ).name + " gesture!", Toast.LENGTH_SHORT
    ).show();
                }
        }
}
```

Now the app will only send a toast if you do a gesture of your choosing. It will be up to you to develop custom actions and reactions to the gestures to integrate them into your app. See Figures 15.7 and 15.8.

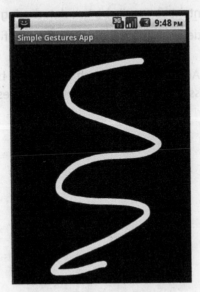

Figure 15.7: *Doing a gesture.*

Figure 15.8: *A toast, or message, when you do a gesture.*

The Least You Need to Know

- Localized resources are organized by country and language code.
- You should have default resources for all content before localizing.
- Nonessential content can easily be translated using Google's translation servers.
- Changes to the device's orientation and other configurations will prompt an Activity restart.
- You can extend Android's multi-touch capabilities by working with the MotionEvent class or by defining your own gestures.

Threads and the Background

In This Chapter

- Introduction to threads and processes
- Background vs. foreground
- Working with AsyncTask
- Building your first service
- Friendly notifications

In day-to-day life, very few of your tasks happen synchronously—that is, one after the other. If you have ever needed a quick lunch but did not have time to collect the ingredients and prepare a meal, you have probably used a microwave to heat a prepared food. The microwave allows you to operate asynchronously. You can pause, start the task, and resume your work confident that the microwave will alert you when your lunch task is ready for consumption.

Java and therefore Android, like most modern languages, has the same concept of separating tasks that need to happen now versus tasks that another agent can execute and complete later. This chapter introduces threads and background services that can be used to process tasks independently from the main task of your app.

What Are Threads?

In the lunch analogy, your workflow is divided into two *threads*. In the first thread, you have started writing some code for your next Android app, you pause for a moment, and you click the Start button on the microwave. This launches the second thread, which enables you to go back to writing some code. Sometime later, the second thread completes and notifies you of the result: hot lunch.

DEFINITION

A **thread** is a unit of processing that is scheduled by an operating system to execute a series of tasks, usually in sequence. Threads are normally members of a process that represents a discrete activity or application.

Most applications as they reach a certain level of complexity need to separate tasks into different threads. It would not be an efficient use of time to force all other tasks to wait for each individual task to complete. Imagine being completely frozen in place while the microwave runs!

There are two primary reasons to branch certain tasks into separate threads. First, you want to allow your app to continue processing essential tasks, like loading the user interface, while other nonessential tasks, like downloading some data from a remote server, run behind. Second, you want your app to be fast and responsive.

Android is particularly concerned about responsiveness, and it has a special dialog prepared to enforce speedy completion. The Application Not Responding (ANR) dialog will display if your app has not responded to the last input within 5 seconds, as shown in Figure 16.1.

Figure 16.1: *The Application Not Responding notice.*

The Android SDK recommends that 100–200 milliseconds is the average user's tolerance for responsiveness, which makes 5 seconds quite generous.

About Processes

Every component of your Android app runs inside a *process*. By default, every app runs in its own process on a single thread. Additional processes can be specified in the Android Manifest. The <application>, <activity>, <provider>, <receiver>, and <service> elements all possess an android:process attribute that can define separate processes in which to run. Specifying an application level process sets this process as the default for all other components.

> **DEFINITION**
>
> A **process** is an instance of a program being executed. A process contains the code to be executed and the activity. Processes in Android are composed of a single, primary thread by default.

Unless otherwise specified, all components are executed in the main thread of their process. From launching the app, clicking a menu icon, responding to a dialog, to entering some data and saving, each of these steps occurs in sequence and blocks the next step. If all activity is executed in the main thread, nothing can happen while waiting for the user to respond to a dialog box.

For this reason, it is best to avoid executing any code in the main thread that will block further user interaction. Android patrols all running processes to optimize memory usage and performance. If processes take too long to complete their tasks, Android considers terminating them.

Processes are evaluated for termination by priority and responsiveness. Lower priority processes are terminated first, followed by higher priority processes as needed. Android prioritizes processes in the following order:

1. **Foreground processes.** An active process that is required for what the user is currently doing.

2. **Visible processes.** An active process that is not in the foreground but affects what is visible on screen. For example, an open, translucent dialog box may be in the foreground, but the Activity screen behind it would still be visible.

3. **Service processes.** An active process. For example, the Service is started, which is not tied to either foreground or visible processes but is still performing desired functions for the user, such as playing music.

4. **Background processes.** An inactive process. The Service is stopped, which is tied to an Activity that is not in the foreground or visible to the user. Many of these processes exist at any given time, and Android prioritizes them by frequency of use. Terminating these processes usually has no visible effect for the user.

5. **Empty processes.** An inactive process primarily used to cache data and speed up an app's start and resume.

The actual ranking your process may receive requires some calculation. Android takes into account the dependencies between processes, rounding up in favor of your process in most cases.

Working with AsyncTask

Writing apps for Android does not mean that every unique operation must complete in under 200 milliseconds or face the ANR Grim Reaper. Rather, think of it as an unspoken contract with your users; you agree not to leave them waiting on an unresponsive screen for more than a second or two before updating them with some new information.

Perhaps the next step in your app's workflow involves uploading a picture to a social networking site. The next step might be to tag this picture with some information, and the upload may logically block this action. Because you have committed to providing your users visual feedback, you need to implement some kind of response within your app, such as a progress bar.

In this case, you need to branch your activity's logic into a separate thread. The main user interface thread will display a progress dialog, and a separate background thread will perform the upload. The Android SDK has implemented the AsyncTask class to handle *asynchronous tasks* and this entire process for you automatically.

DEFINITION

An **asynchronous task** executes on a background thread, separate from the main user interface thread of an app. The asynchronous thread needs to periodically update the main thread with the results of its progress.

In this case, the AsyncTask class would be:

```
private class MyAsyncTask extends AsyncTask<URL,Int,String>{}
```

It must be instanced as a subclass to be used, and the class takes three types:

- **Params.** The type of parameters provided to the task to execute. In this case, the type is of URL.

- **Progress.** The type of units in which to publish progress information. In this case, the type is of Integer.

- **Result.** The type of the result to return to the main thread. In this case, type is of String.

Not all types are required, but you must pass Void if you do not intend to use a particular type. Calling an AsyncTask subclass instructs Android to launch the task in a new background thread and manage the entire thread life cycle for you. This involves executing the following methods:

- **onPreExecute().** This executes on the main, user interface thread immediately upon calling AsyncTask. Use this method to set up the task and render any interface elements, such as a progress bar, that you wish to be updated.

- **doInBackground().** This method takes the Params type and is called as soon as onPreExecute() completes. You will execute all of your expensive or long-running code here.

- **publishProgress().** This method takes the Progress type and can be called from within doInBackground(). Call this method at frequent intervals to update the progress of your task.

- **onProgressUpdate().** This method also takes the Progress type and is called by publishProgress(). Use this method to provide updates to the user interface thread, such as incrementing the progress bar.

- **onPostExecute().** This method takes the Result type and is executed as soon as doInBackground() has completed.

Now, you can build a simple asynchronous task. In this example, you design a simple Activity that launches a background thread to count to 100 while keeping the user interface thread up to date with a ProgressDialog:

```java
import android.app.Activity;
import android.app.ProgressDialog;
import android.content.Context;
import android.content.DialogInterface;
import android.os.AsyncTask;
import android.os.Bundle;
import android.widget.TextView;
public class MyAsyncActivity extends Activity {
        protected ProgressDialog _progressDialog;
        protected TextView _resultField;
        protected DoAsyncTask _DoAsyncTask;

        @Override
        public void onCreate( Bundle savedInstanceState )
        {

                super.onCreate( savedInstanceState );

                setContentView( R.layout.at );

                resultField = (TextView) this.findViewById( R.id.result_
        field );

                progressDialog = new ProgressDialog( this ;
                progressDialog.setProgressStyle( ProgressDialog.STYLE_
        HORIZONTAL );
                progressDialog.setTitle( "Async Task Progress" );
                progressDialog.setMax( 100 );
                progressDialog.setCancelable( true );
                progressDialog.setOnCancelListener( new
        CancelButtonListener() );

                DoAsyncTask = new DoAsyncTask();
                DoAsyncTask.execute( this );
        }

        protected class CancelButtonListener implements
        DialogInterface.OnCancelListener
        {

                @Override
                public void onCancel( DialogInterface dialog ) {
                        DoAsyncTask.cancel(true);
                }
        }

        protected class DoAsyncTask extends AsyncTask<Context,
        Integer, String>
        {
```

```java
            @Override
            protected void onPreExecute()
            {
                    super.onPreExecute();
                    progressDialog.setMessage( "Your AsyncTask is
➥Processing..." );
                    progressDialog.show();
            }

            @Override
            protected String doInBackground( Context... params )
            {
                    for( int i = 0; i<=100; i++ )
                    {
                            try{
                                    Thread.sleep( 50 );
                                    publishProgress( i );
                                    i++;
                            }
                            catch( Exception e ){}
                    }
                    return "Your AsyncTask Has Completed!";
            }

            @Override
            protected void onProgressUpdate( Integer... values )
            {
                    super.onProgressUpdate( values );
                    progressDialog.setProgress( values[0] );
            }
            @Override
            protected void onPostExecute( String result )
            {
                    super.onPostExecute(result);
                    progressDialog.dismiss();
                    resultField.setText( result );
            }
            @Override
            protected void onCancelled()
            {
                    super.onCancelled();
                    resultField.setText( "You cancelled the
➥AsyncTask!" );
            }
        }
}
```

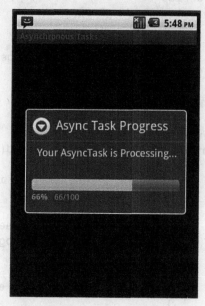

Figure 16.2: *The AsyncTask progress bar.*

With AsyncTask, Android handles all of the work of managing application, process, and thread resources, which allows you to program faster and more efficiently.

Dispatching Threads

While Android's automated background thread management with AsyncTask is frequently convenient and useful, you will sometimes want to manage threads on your own. Thread management is its own field of study and requires more research and understanding about thread safety, memory leaks, and concurrent, asynchronous development.

 GOOGLE IT

The Android Developer's Blog posts detailed guides for accomplishing complex tasks through the Android SDK. Bookmark http://android-developers.blogspot.com and visit weekly for valuable insight into Android programming.

Thread management in Android begins with the Thread class. A thread can be instanced from the Thread class, in which case you provide a Runnable object that can execute your task. You can also create a new class that extends Thread, in which case you need to override the run() method. Both are initiated by calling the start() method.

The whole scope of thread management is encompassed in these classes:

- **Thread.** A unit of execution. Each thread has a priority, which can be set by calling setPriority().

- **MessageQueue.** Holds the list of messages dispatched, normally managed by the Handler.

- **Handler.** A mechanism for sending Messages and processing Runnables. Each Handler is tied to a single Thread and a single MessageQueue.

- **Runnable.** A command to be executed on a Thread.

- **Message.** Contains a description and content, which can hold any arbitrary data.

- **ThreadGroup.** Organizes threads and thread groups into a hierarchical structure. Each child thread or group has a parent ThreadGroup.

Using Threads, Handlers, and Messages, you can refactor the AsyncTask example from previously to use self-managed threads:

```
import android.app.Activity;
import android.app.ProgressDialog;
import android.content.DialogInterface;
import android.os.Bundle;
import android.os.Handler;
import android.os.Message;
import android.widget.TextView;
public class ThreadedTask extends Activity implements Runnable {
        protected ProgressDialog _progressDialog;
        protected TextView _resultField;
        protected Thread thread;

        @Override
        public void onCreate( Bundle savedInstanceState )
        {
                super.onCreate(savedInstanceState);
```

continues

```
              setContentView( R.layout.at );

              resultField = (TextView) this.findViewById(R.id.result_
    field);

              progressDialog = new ProgressDialog(this);
              progressDialog.setProgressStyle(ProgressDialog.STYLE_
    HORIZONTAL);

              progressDialog.setTitle( "Async Task Progress" );
              progressDialog.setMax( 100 );
              progressDialog.setCancelable( true );
              progressDialog.setOnCancelListener( new
    CancelButtonListener() );
              progressDialog.setMessage( "Your AsyncTask is
    Processing..." );
              progressDialog.show();
              thread = new Thread( this );
              thread.start();
        }

        protected class CancelButtonListener implements
    DialogInterface.OnCancelListener
        {
              @Override
              public void onCancel( DialogInterface dialog ) {
                    thread.stop();
              }
        }

        public void run() {
              for( int i = 0; i<=100; i++ )
              {
                    try{
                          Thread.sleep( 50 );
                          i++;
                          progressDialog.setProgress( i );
                    }
                    catch( Exception e )
                    {
                          handler.sendEmptyMessage( RESULT_CANCELED
    );
                    }
              }
              handler.sendEmptyMessage( RESULT_OK );
        }
```

```
private Handler handler = new Handler() {
    @Override
    public void handleMessage(Message msg) {
        if( RESULT_OK == msg.what )
        {
            progressDialog.dismiss();
            resultField.setText( "This Background
Task Completed!" );
        }
        else
        {
            progressDialog.dismiss();
            resultField.setText( "This Background
Task Did Not Complete!" );
        }
    }
};
}
```

Running this code in the emulator, the output is identical; but as you will discover, the complexity of managing threads on your own increases substantially.

Services

On desktop class operating systems, a service frequently references an application that runs in the background or is not visible to the user. In this context, services frequently run independent from an application, though they might compliment an application in some way.

On Android, this is not the case. Services run in the same process as the app on the main user interface thread. By creating and using a service, you do not escape the 5-second requirements for app responsiveness. Rather, services provide two core functions to the Android app:

- Services provide a mechanism to continue executing some task, like playing media, while the user is not actively interacting with the app.

- Services provide a way to communicate information with other apps on the device.

Services adhere to the same thread rules as Activities. If your service needs to execute blocking or long-running tasks, these should be executed in their own threads, just as you would within an Activity.

The Service Life Cycle

Just as with Activities, you first declare a Service in the AndroidManifest.xml using the <service> tag. A Service begins its life cycle either by being started or bound. Calling startService() launches the Service for immediate execution, while calling bindService() creates the Service for longer connection periods.

The Service life cycle is similar to the Activity life cycle, which you explored in Chapter 5, but far more straightforward. The entire Service life cycle occurs between onCreate() and onDestroy(). The active life cycle occurs after onStart() if the Service is started, or between onBind() and onUnBind() if the Service is bound. There is no onStop() event; the Service passes immediately to onDestroy().

If you intend for your Service to be both started and bindable, both life cycles can occur, and you should plan your code accordingly. If the system is running low on memory, Android will evaluate your Services for termination based on the status of their corresponding foreground process as well as against the priority of any processes that have bound to the Service.

Building a Basic Service

The thought of building a Service seems intimidating, probably conjuring discomforting images of black box behavior. In Android development, they are very similar to building Activities. You can visualize the steps as follows:

1. Your Service will be a Java class, so declare it first in AndroidManifest.xml.

2. Create the class in Eclipse, and extend the Service class.

3. Implement placeholder life cycle methods for onCreate(), onStart(), and onDestroy().

4. Start and stop the Service from an Activity.

With these guidelines, you can create a simple media player service. This will be a barebones, proof-of-concept. Your service will send a toast notification with each change in its life cycle. You will instance a MediaPlayer object, load a music file of your choice, and start playing the file onStart(). To wrap it up—onDestroy(); you will stop playing the music.

Your new Service class as detailed:

```
import android.app.Service;
import android.content.Intent;
import android.media.MediaPlayer;
import android.os.IBinder;
import android.widget.Toast;
public class MediaPlayerService extends Service {
        private MediaPlayer _MediaPlayer;
        public static boolean isStarted = false;

        @Override
        public void onCreate() {
                Toast.makeText( this, "MediaPlayerService Created.
Music Loaded.", Toast.LENGTH_LONG ).show();
                MediaPlayer = MediaPlayer.create( this, R.raw.your_
music );
        }
        @Override
        public void onStart( Intent intent, Int int) {
                Toast.makeText( this, "MediaPlayerService Started.
Music Started.", Toast.LENGTH_LONG ).show();
                isStarted = true;
                MediaPlayer.start();
        }

        @Override
        public IBinder onBind( Intent intent ) {
                return null;
        }

        @Override
        public void onDestroy() {
                Toast.makeText( this, "MediaPlayerService Stopped.
Music Stopped.", Toast.LENGTH_LONG ).show();
                isStarted = false;
                MediaPlayer.stop();
        }
}
```

Congratulations! You have just implemented your first Android Service. While you can retrieve the active state of your Service from the OS, it is sometimes simpler to keep track of this yourself. Using the Boolean isStarted, you can keep account of your Service's running state. You will likely want to use this informal tracking system only for early development and testing, and rely upon the OS's authoritative answers for your prime time release.

Once inside your app's project, you can start and stop this Service from any of your app's activities. You may want the Service to start with your app and stop onPause(), or you may want the Service to run only in the context of a limited number of activities. For starters, implement the Service in a new Activity based on a simple button.

```java
import android.app.Activity;
import android.content.Intent;
import android.os.Bundle;
import android.view.View;
import android.view.View.OnClickListener;
import android.widget.Button;
public class PlayMusic extends Activity implements OnClickListener {

      Button playTheMusic;
      @Override
      public void onCreate( Bundle savedInstanceState ) {
            super.onCreate( savedInstanceState );
            setContentView( R.layout.main );

            playTheMusic = (Button)findViewById( R.id.buttonControl
);
            playTheMusic.setOnClickListener( this );
            playTheMusic.setText( "Play" );
      }
      public void onClick( View view ) {
            switch (view.getId()) {
                  case R.id.buttonControl:
                  if( MediaPlayerService.isStarted )
                  {
                        stopService( new Intent( this,
MediaPlayerService.class ) );
                        playTheMusic.setText( "Play" );
                  }
                  else
                  {
                        startService( new Intent( this,
MediaPlayerService.class ) );
                        playTheMusic.setText( "Stop" );
                  }
```

```
                                    break;
                        }
                }
        }
```

While a simple block of code, you can imagine ways in which this can be expanded. From creating your own live wallpaper, screensaver, or media manager to building a pedometer, services can enable you to extend your app in exciting ways.

Using Notifications

As a general social rule, people generally prefer the least intrusive interruptions necessary to alert them of some news. While running and shouting into a co-worker's office may be appropriate on the day of Christmas bonuses, it probably is not needed to ask a question about that last TPS report.

The same rules apply on mobile devices. Although toast alerts are extremely useful during development and for testing and debugging, they are the social equivalent of yelling, "Fire!" in a movie theater. Only use them in your Market app when you really need to.

For a more elegant and nonintrusive method of prompting your users, "Pardon me, I have something that requires your attention … at your convenience", Android offers the notification layer. Any Activity or Service can submit a Notification to the NotificationManager, which places the alert into the user's status bar.

ANDROID DOES

The Status Bar, located at the top of the Android screen, contains a list of all notifications, from missed calls to text messages. These notifications consist of a message pertaining to what happened, and an Intent that is fired when the user touches the notification.

Services are best equipped to handle notifications, as they can be running while the foreground Activity is not in the foreground. To guarantee that your app and your users receive notifications, implement them in your app's services.

Submitting a notification follows these steps:

1. Obtain a reference to the NotificationManager, a system service.

2. Instance a Notification, with an icon to display in the status bar, a text to display, and the time of the notification.

3. Give the Notification some content: a name, description, and Intent.

4. Pass the finished Notification back to the NotificationManager reference.

Here's the code for all four steps.

```
//Reference the NotificationManager
NotificationManager _NotificationManager = (NotificationManager)
getSystemService( Context.NOTIFICATION_SERVICE );

//Instance a notification
int notificationIcon = R.drawable.icon;
CharSequence notificationText = "Your Attention is Requested";
long notificationTime = System.currentTimeMillis();
Notification notification = new Notification( notificationIcon,
notificationText, notificationTime );

//Populate the notification with context and cotent
Context context = getApplicationContext();
CharSequence contentTitle = "New Content for Your Review";
CharSequence contentText = "You have new content available in
my app.";
Intent notificationIntent = new Intent( this, PlayMusic.class
);
PendingIntent contentIntent = PendingIntent.getActivity( this,
0, notificationIntent, 0 );
notification.setLatestEventInfo( context, contentTitle,
contentText, contentIntent );

//Pass the notification back to the NotificationManager
_NotificationManager.notify( 1, notification );
```

If you were developing a weather application, you might want to submit notifications in response to storm warnings. Of course, as the weather changes, your users would thank you not to submit new notifications, but to amend the old. You can update your existing notifications by calling setLatestEventInfo() and issuing notify() to the NotificationManager again.

In most cases, the default notification settings will be suitable for your needs, but you may consider the accessibility needs of your users. This is the notification default code:

```
notification.defaults |= Notification.DEFAULT_SOUND
```

By setting notification defaults, you can add sounds, vibration, and lights to your notifications. Here are the pieces:

- **DEFAULT_ALL.** Adds all default elements to the notification.

- **DEFAULT_SOUND.** Adds the system default notification sound.

- **DEFAULT_VIBRATE.** Adds the system default vibration setting.

- **DEFAULT_LIGHTS.** Adds the system default LED light alert.

You can override the default values and settings with your own sounds, unique vibration patterns, and light signals. You will likely want to make all of these settings configurable within your app.

The Least You Need to Know

- The Application Not Responsive will display if your user interface does not update within 5 seconds of an action.
- AsyncTask provides an auto-managed process for issuing tasks to background threads.
- Manual thread management is possible through the Thread and Handler classes.
- Services allow tasks to run while your app is not in focus.
- Use Notifications to alert your users of important information.

SQL and Databases

In This Chapter

- Using SQLite in Android
- Working with Structured Query Language
- Creating and upgrading databases
- Storing and finding information
- Displaying database content in your app

Whether you hope your next app will solve world problems or entertain young children, you will need to store, find, and update information that your users provide through using your app. Some of this content, like options and preferences, you can easily store and access through SharedPreferences, which you used to store basic content in Chapter 8.

As your app scales, you will want to store more data, access it faster, and update it quickly. For this purpose, you need a true database to manage this information. This chapter introduces SQLite, a powerful, lightweight, and public domain database integrated directly into Android.

SQL at a Glance

A conventional database is a collection of tables that are represented as rows and columns. Tables are often visualized as spreadsheets, but while spreadsheets are optimized for displaying and editing data in a user interface, databases are optimized for storing data and accessing it quickly.

Communication with a database occurs through *Structured Query Language (SQL)*. Android provides a helper class and methods, allowing you to interact with the SQLite database without using SQL. However, it is important to understand that these convenience functions simply convert the code into SQL, which is passed to the database.

> **DEFINITION**
>
> **SQL** is the most widely used language for communicating with databases, and it is used to query, insert, update, and delete data. SQL also has the power to change the structure of the database, modifying tables, columns, and permissions.

You need to understand some of the syntax of SQL both to make the most of the convenience functions and to be able to write your own SQL for communicating directly with the database. The simplest structure in a database is a table. As you develop a specification for your database structure, you will likely need several.

Let's create a table with a list of recipes and their corresponding types of cuisines:

Basic Table in SQLite

_id (integer primary key)	recipe_name (text)	cuisine (text)
1	Artichoke Hearts	Mediterranean
2	French Fries	American
3	Jambalaya	Cajun
4	Apple Pie	American

Creating and modifying tables typically happens only when your app is first installed or when you release updates. You will use these SQL commands less frequently, but they are important; your tables cannot exist without them.

To create the previous table, issue a create table statement:

```
create table recipes (
id integer primary key autoincrement,
recipe_name text,
cuisine text );
```

This creates an empty table with three columns. A core concept of relational databases is that each table has a primary key, which allows each record to be uniquely identified. This allows the database to know, unambiguously, that recipes._id = 3 is Jambalaya. The name _id is chosen here because it has special meaning for Android, not because it matters to SQLite.

GOOGLE IT

Unlike most other databases, the field type you specify for your columns, such as text, datetime, or numeric, are just guidelines. SQLite does not enforce these types, and you can insert any value into any field—with the exception of autoincrement, which is special.

With the table in place, you might later realize that you need more columns. You can add a created_date with the alter table statement:

```
alter table recipes
add created_date datetime;
```

If you ever need to delete a table, do so by issuing the drop table statement:

```
drop table recipes;
```

Most of your interaction with the database will take the form of queries or data manipulation. This is accomplished with four statements:

- **SELECT.** Returns a result set of rows and columns based on a query. For instance, here is a short code that gives a result for "American" cuisine:

```
select recipe_name
from recipes
where cuisine='American';
```

- **INSERT.** Adds rows to a table. For instance, the code below adds a row with the values "French Fries" and "Spanish" (we will correct the cuisine conflict in a minute).

```
insert into recipes( _id, recipe_name, cuisine )
values( null, 'French Fries', 'Spanish' );
```

- **UPDATE.** Updates specified rows within a table. In this case, in the "French Fries" row, the cuisine text is changed to "American".

```
update recipes
set cuisine='American'
where recipe_name='French Fries';
```

- **DELETE.** Deletes specified rows from a table. With a short code, "Spanish" is removed from the cuisine list.

```
delete from recipes
where cuisine='Spanish';
```

SQLite's use of SQL, unlike Java, is case-insensitive for the query structure. This means that "SELECT FROM" and "select from" are treated identically. Text values stored in the database do have case sensitivity when using mathematical operators, like = or !=. Operators such as "like" will be case-insensitive.

Your table is still empty, so let's add some content to it.

```
insert into recipes( _id, recipe_name, cuisine )
      values( null, 'Artichoke Hearts', 'Mediteranian' );
insert into recipes( _id, recipe_name, cuisine )
      values( null, 'French Fries', 'American' );
insert into recipes( _id, recipe_name, cuisine )
      values( null, 'Apple Pie', 'American' );
```

If you ran the following:

```
select * from recipes where cuisine='american';
```

You would see no results, while executing the following:

```
select * from recipes where cuisine like 'american';
```

This would return both "French Fries" and "Apple Pie".

Writing SQL is its own art form and requires study and practice. W3Schools.com and SQL.org provide excellent and free tutorials on working with SQL in depth.

Intro to SQLite

Working with databases is like driving a car. Once you know how to drive, you can get behind the wheel of almost any car and know how it works. But every car is different and has its own quirks, which means you have to spend a little time getting to know each other before you are comfortable.

SQLite is a database designed to be lightweight and efficient. As compiled for Android, its total size is less than 250 KB, which makes it one of the smallest database server applications available. A SQLite database also encompasses just a single file, which makes the application fast, flexible, and easy to move.

Android stores the database file in /data/<your package name>/databases. You can copy this file to the external SD card to make it available to other apps. Working in debug or DDMS mode, you can directly view and query this database.

Building the Helper Class

Android has abstracted some database interactions by implementing a helper class, SQLiteOpenHelper. You must use this class to create and upgrade your database. The class takes three methods:

- **onCreate().** Creates the database if it does not already exist.

- **onUpgrade().** Runs if the client has an older version of the database installed.

- **onOpen().** Optional method. Runs when the database is opened.

You probably want to build a larger helper class to manage all of your database interactions, and you can nest this class inside your more robust helper class:

```
import android.content.ContentValues;
import android.content.Context;
import android.database.Cursor;
import android.database.SQLException;
import android.database.sqlite.SQLiteDatabase;
import android.database.sqlite.SQLiteOpenHelper;
import android.database.sqlite.SQLiteStatement;

public class DBHelper  {
    private static final String _DatabaseName = "recipes.db";
    private static final int _DatabaseVersion = 2;
    private static final String _RecipesTable = "recipes";

    private static final String _IdColumn = "_id";
    private static final String _RecipeColumn = "recipe";
    private static final String _CuisineColumn = "cuisine";
    private static final String _CreatedColumn = "created_date";

    private Context _Context;
    private SQLiteDatabase db;
    private DBOpenHelper _DBOpenHelper;

    public DBHelper(Context context) {
        this._Context = context;
```

continues

```
        DBOpenHelper = new DBOpenHelper(this._Context);
    }

    private static class DBOpenHelper extends SQLiteOpenHelper {

        DBOpenHelper(Context context) {
            super(context, _DatabaseName, null, _
DatabaseVersion);
        }

        String createTable = "CREATE TABLE " + _RecipesTable +
        "( " + _IdColumn + " INTEGER PRIMARY KEY autoincrement,
" +
        RecipeColumn + " TEXT," +
        CuisineColumn + " TEXT," +
        CreatedColumn + " DATETIME );";
        String upgradeTable = "ALTER TABKE " + _RecipesTable +
        " ADD " + _CreatedColumn + " DATETIME;";
        String dropTable = "DROP TABLE " + _RecipesTable + ";";

        ContentValues _DefaultContent = new ContentValues();

        @Override
        public void onCreate(SQLiteDatabase db) {
            db.execSQL( createTable );
            DefaultContent.put( _RecipeColumn, "Apple Pie"
);
            DefaultContent.put( _CuisineColumn, "American"
);
            DefaultContent.put( _CreatedColumn, System.
currentTimeMillis() );
            db.insert( _RecipesTable, null, _DefaultContent
);
        }

        @Override
        public void onUpgrade(SQLiteDatabase db, int
oldVersion, int newVersion) {
            if( 1 == oldVersion && 2 == newVersion )
                db.execSQL( upgradeTable );
            else
            {
                db.execSQL( dropTable );
                onCreate(db);
            }
```

```
        }
      }
    }
```

With the DBHelper class in place, it can be instanced anywhere else in your application. Because the DBHelper constructor instances DBOpenHelper, you will have a valid database object to work with. DBHelper needs to be extended to open and close connections with the database and provide other convenience functions for the rest of your app.

PITFALL

While the values for table name, column names, and version are hard-coded in this example, you will want to store these as in a lookup table in the database or elsewhere in your app's metadata.

Database Interaction

With the groundwork laid by SQLiteOpenHelper, you can begin working with the SQLiteDatabase class to query and modify data. The SQLiteDatabase class abstracts much of the logistics of managing database connections and interpreting results. As such, it has a few nuances worth considering.

1. Query results are returned as a *cursor*, whether you write your own raw SQL with the rawQuery() method or use the convenience query() method.

2. Writing data to an SQLite database is slower than on other database platforms.

3. Because the SQLite database is a single file, it is important to use transactions when updating and inserting many records at a time. The beginTransaction(), setTransactionSuccessful(), endTransaction() methods will ensure your entire SQL tasks complete or fail, allowing you to try again. This prevents painful data inconsistencies.

DEFINITION

Like an array, a **cursor** contains an arbitrary number of objects, which can be iterated. Unlike an array, a cursor maintains a relationship with the query that generated it, enabling the cursor to request a requery of its results and other operations.

Before diving in and extending the DBHelper class, take a moment to review some of the most common and useful methods of the SQLiteDatabase class in the following table.

Common Methods of the SQLiteDatabase Class

Method	Purpose	Returns
close()	Closes a connection to the database	Nothing
isOpen()	Checks if the database is opened	True, if open
isReadOnly()	Checks if the database is in a read-only state	True, if the database is open and read-only
insert()	Inserts a new row into the database	The row id of the new record
query()	Submits a query to the database	A cursor of the result set
rawQuery()	Submits formed SQL statements to the database	A cursor of the result set
replace()	Replaces a row in a table with an insert	The row id of the record
update()	Updates rows in a table with new values	The number of rows affected

Now, these methods can be implemented into the DBHelper class, after the DBOpenHelper class closes, for use anywhere within the app:

```
private String[] RecipeColumns = new String[]{_IdColumn,_
    ➥RecipeColumn,_CuisineColumn,_CreatedColumn};

public DBHelper open() throws SQLException
{
    db = _DBOpenHelper.getWritableDatabase();
    return this;
}
public void close()
{
    DBOpenHelper.close();
}

public long addRecipe(String recipe, String cuisine)
{
    ContentValues values = new ContentValues();
```

```
        values.put( _CreatedColumn, recipe );
        values.put( _CreatedColumn, cuisine );
        values.put( _CreatedColumn, System.currentTimeMillis() );
        return db.insert( _RecipesTable, null, values );
}

public boolean deleteRecipe(long rowId)
{
        return db.delete( _RecipesTable, _IdColumn + "=" + rowId, null
) > 0;
}

public Cursor getAllRecipes()
{
        return db.query( _RecipesTable,
        RecipeColumns,
        null,
        null,
        null,
        null,
        null );
}

public Cursor getRecipe(long rowId) throws SQLException
{
        Cursor cursor =
        db.query( true,
        RecipesTable,
        RecipeColumns,
        IdColumn + "=" + rowId,
        null,
        null,
        null,
        null,
        null );
        if ( cursor != null ) {
                cursor.moveToFirst();
        }
        return cursor;
}

public boolean updateRecipe(long rowId, String recipe, String
cuisine, Date created )
{
        ContentValues values = new ContentValues();
        values.put( _RecipeColumn, recipe );
        values.put( _CuisineColumn, cuisine );
        values.put( _CreatedColumn, "created" );
```

continues

```
        return db.update( _RecipesTable, values, _IdColumn + "=" +
    ➥rowId, null ) > 0;
}
```

You now have a set of seven methods—open(), close(), addRecipe(), deleteRecipe(), getAllRecipes(), getRecipe(), and updateRecipe()—which you can conveniently call to instantly execute these actions on the database. You will likely want more than these, and there will be times when you need to interact directly with the database object. Using these as a starting point, you can easily build your own and modify them to your specific needs.

More SQL Detail

From the convenience functions you just created, you may have noticed that the query() method in particular requires quite a few parameters, for most of which you used null. As Android must convert all database interactions into valid SQL statements before passing them to the database, the methods on the SQLiteDatabase class provide you a way to modify the SQL without using true SQL.

The query() method corresponds to the SQL Select statement, which is arguably the most sophisticated statement in SQL. A Select statement has six primary clauses, which can be used to refine what data is returned and how that data is formatted. Not all clauses are required, but they must be listed in order:

1. **SELECT.** Required. Defines which columns to return in the result set. "*" indicates all columns should be included.

2. **FROM.** Required. Defines which table or tables to perform the select on. Can be used in conjunction with a JOIN clause to specify relationships between tables.

3. **WHERE.** Limiting conditions to apply to the select in order to reduce the set of results to specific criteria.

4. **GROUP BY.** Specifies repeating values to group together. Required if using aggregate functions like COUNT() or SUM().

5. **HAVING.** Used only with a GROUP BY clause. Used to limit the result set based on how values are grouped.

6. **ORDER BY.** Specifies ascending or descending order by column for the result set.

The query() method is defined as:

```
query (String table, String[] columns, String selection, String[]
  ⮞selectionArgs, String groupBy, String having, String orderBy)
```

In the context of the Select statement, this method's parameters align with the clauses of the Select as follows:

- **table** matches the FROM clause.
- **columns** matches the SELECT clause.
- **selection** matches the WHERE clause.
- **selectionArgs** provides additional flexibility for the WHERE clause.
- **groupBy**, **having**, and **orderBy** match the corresponding clauses.

Bring the Data into View

In the DBOpenHelper class, you added a single, default insert to the recipes database onCreate(). You can add more default records to help you fully test the next step, which is to display the contents of the database inside an Activity. This process is called *data binding*.

> **DEFINITION**
>
> **Data binding** is the process of joining two separate information sources together, frequently between a data source and XML or a user interface.

To demonstrate a proof-of-concept, you can create a simple LinearView layout with a single Spinner widget. Using DBHelper, query the database for all recipes and fill the Spinner control with the available recipe names. In order to do this, you will need to bind the Cursor object with the View.

Android simplifies the data binding process by providing adapters to bridge a data source with a View object. The SimpleCursorAdapter class provides a simple way to achieve this:

```
import android.app.Activity;
import android.database.Cursor;
import android.os.Bundle;
import android.view.View;
```

continues

```
import android.widget.SimpleCursorAdapter;
import android.widget.Spinner;
public class DisplayRecipes extends Activity {

        DBHelper _DBHelper;
        @Override
        public void onCreate( Bundle savedInstanceState ) {
                super.onCreate( savedInstanceState );
                setContentView( R.layout.main );

                _DBHelper = new DBHelper( this );
                _DBHelper.open();

                Spinner recipeList = (Spinner) findViewById( R.id.
recipeList );
                Cursor cur = _DBHelper.getAllRecipes();

                SimpleCursorAdapter adapter = new SimpleCursorAdapter(
this,
                android.R.layout.simple_spinner_item,
                cur,
                new String[] {"recipe"},
                new int[] {android.R.id.text1} );

                adapter.setDropDownViewResource( android.R.layout.
simple_spinner_dropdown_item );
                recipeList.setAdapter( adapter );
        }

}
```

Run this in the emulator, and you will see a simple Spinner pick list with the recipe names you have defined in the database. The SimpleCursorAdapter class constructor takes five parameters:

- **Context.** Usually this or the context from an Activity.

- **Layout.** The layout template used to associate data rows and columns with TextView or ImageView objects.

- **Cursor.** A cursor representing some query.

- **From.** A String array of column names returned within the cursor.

- **To.** An Integer array of View element ids to match against the From array, in the same order.

From the previous Spinner example, notice that Android layout templates were used instead of templates created in your own resource directory. This is perfectly acceptable for testing functionality, but you will likely want to take control and fashion your own templates. Using the same cursor, you can now bind the data to a List.

First, modify main.xml to include only two child view elements: a ListView and a TextView. Android reserves certain View ids for special purposes, which are designated by the @android:id tag. The ListView uses the reserved id list, which instructs Android that this ListView is special and will be inflated with some other content. The TextView uses the reserved id empty, which instructs Android what to display if the ListView is empty.

Let's look at the first step, adding the ListView and the TextView:

```xml
<?xml version="1.0" encoding="utf-8"?>
<LinearLayout
        xmlns:android="http://schemas.android.com/apk/res/android"
        android:layout_width="fill_parent"
        android:layout_height="fill_parent"
>
<ListView
        android:id="@android:id/list"
        android:layout_width="wrap_content"
        android:layout_height="wrap_content"
/>
<TextView
        android:id="@android:id/empty"
        android:layout_width="wrap_content"
        android:layout_height="wrap_content"
        android:text="No Recipes in the Database."
/>
</LinearLayout>
```

Now, you need a layout template to format the inflated results from the cursor. This should define a parent ViewGroup with TextView and ImageView child elements to bind with data.

```xml
<LinearLayout
xmlns:android="http://schemas.android.com/apk/res/android"
        android:layout_width="fill_parent"
        android:layout_height="fill_parent"
        android:orientation="horizontal"
        android:padding="15dp">
        <TextView
```

continues

```
                    android:id="@+id/_id"
                    android:layout_width="0dp"
                    android:layout_height="0dp"
                    android:visibility="invisible"
            />
            <TextView
                    android:id="@+id/recipe"
                    android:layout_width="wrap_content"
                    android:layout_height="wrap_content"
            />
            <TextView
                    android:id="@+id/recipe_pad"
                    android:layout_width="wrap_content"
                    android:layout_height="wrap_content"
                    android:text=" - "
            />
            <TextView
                    android:id="@+id/cuisine"
                    android:layout_width="wrap_content"
                    android:layout_height="wrap_content"
            />
            <TextView
                    android:id="@+id/cuisine_pad"
                    android:layout_width="wrap_content"
                    android:layout_height="wrap_content"
                    android:text=" - "
            />
            <TextView
                    android:id="@+id/created"
                    android:layout_width="wrap_content"
                    android:layout_height="wrap_content"
            />
    </LinearLayout>
```

Each TextView element corresponds either to a field in the database or some formatting used to improve the aesthetic of the list. You likely want to format this into ordered rows and columns with column headers, but for now this provides a simple display of content.

You notice that _id is included, but invisible. If you were to later implement a click or touch listener to the list, having the recipe id immediately available in the View content can save you some time formulating a query. With the two layout resources in place, you need only adapt the SimpleCursorAdapter in DisplayRecipes to fill the List instead of the Spinner.

```
public class DisplayRecipes extends ListActivity {

        DBHelper _DBHelper;
        @Override
        public void onCreate( Bundle savedInstanceState ) {
                super.onCreate( savedInstanceState );
                setContentView( R.layout.main );

                DBHelper = new DBHelper( this );
                DBHelper.open();

                Cursor cur = _DBHelper.getAllRecipes();

                SimpleCursorAdapter adapter = new SimpleCursorAdapter(
    ➥this,
                R.layout.recipes,
                cur,
                new String[] {"_id","recipe","cuisine","created_date"},
                new int[] {R.id._id, R.id.recipe, R.id.cuisine, R.id.
    ➥created} );
                setListAdapter( adapter );
        }

}
```

From here, you can implement the fruits of your database labor throughout your app. By creating a few generic layout templates and some database helper functions, you can immediately expose the contents of your database with just a few lines of code.

Additional SQLite Tools

As you develop your app, the first obstacle you are likely to reach is the limitations of the SQLiteDatabase classes helper functions. While fast and easy, if your database grows beyond a few tables, you will need to run increasingly complex queries. You always have the option to do this by writing your own SQL, but Eclipse and the ADT plugin will not capture errors in your code. It is cheaper to catch errors at compile time than by hitting fatal error pages in the app.

The Android SDK alleviates some of this tension with the SQLiteQueryBuilder class. This class provides methods to add additional depth or specificity to a query:

- **appendColumns().** Adds a list of column names to a query.

- **appendWhere().** Adds an additional WHERE condition to a query.

- **buildQueryString().** Produces an SQL SELECT string for use as a query.

- **buildUnionQuery().** Produces a UNION query of one or more SELECT statements.

There will also be times in the course of developing, testing, and debugging your app that you need more immediate access to the database. It would be extremely inefficient to have to write execute query() methods just to inspect your database.

You have a variety of options for accessing the database from the emulator or your Android device and inspecting it directly. Here are two popular options:

1. Using the Android Debug Bridge (ADB), which you learn about in Chapter 19, you can connect directly to the database from a console connection and issue commands.

2. Using the DDMS perspective in Eclipse, you can copy the SQLite database to your local machine.

   a. Install SQLite for your operating system from http://www.sqlite.org and inspect the downloaded database directly.

   b. Install the SQLite Manager for Firefox, and inspect any SQLite database directly from the browser. http://sqlite-manager.googlecode.com.

   c. Install an application like SQLite Browser, http://sqlitebrowser. sourceforge.net, and inspect the database from a desktop application.

Beyond these free options, a vast array of commercial offerings are also available, though you will likely find all of the resources you need through the Android Developer site and Google.

Apps large and small rely on databases to store and find information. Once you have implemented your database, you will find it increasingly easy to work with information. Having data stored in the database opens other opportunities for your application, such as sharing information between apps.

The Least You Need to Know

- Android has directly integrated the SQLite database into the OS.
- Databases are created, upgraded, and instanced through the SQLiteOpenHelper class.
- All database communication occurs in SQL.
- Convenience methods provide an easy way to convert Java into SQL.
- Data binding is the process of associating the results of a query with View objects on screen.

ContentProviders

In This Chapter

- Intro to ContentProviders
- Sharing data between apps
- Querying other apps
- Creating your own provider

Imagine there was no central point of access to a contacts database. It would then be up to each individual app to create and maintain its own set of contacts. Users could quickly find themselves lost, forced to juggle through multiple apps just to find a phone number.

Android solves this problem with an elegant solution: ContentProviders. Providers enable apps to publish information they collect to other apps. This chapter explores how to use Providers to pull useful information from other apps into your own, as well as how to make your valuable information available to others.

Introduction to ContentProviders

Each Android app receives its own unique Linux user identification and data directory in /data/data/<Your Package Name>. This directory is secured, preventing all apps but your own from accessing it. If you think of the financial and banking apps you may have on your device, you can quickly see why this is a good policy.

In order for an app's content to be considered public or accessible to other apps, the app must supply a *ContentProvider*.

Many of the Android core apps already supply Providers, some of which you have already used with Intents. Here is a table with the most common ones:

Common Android ContentProviders

Provider	Included Data
Browser history	Bookmarks and browsing
CallLog	Recent and missed calls
ContactsContract	Contacts database
MediaStore	Image, audio, and video files in the MediaStore index
Settings	Device and OS settings
UserDictionary	Words added by the user

Provider Syntax

You may recall from Chapter 5 that Android allows certain undefined Intents to be called against a URI, such as dialing a contact:

```
startActivity( new Intent( "android.intent.action.DIAL",
Uri.parse( "content://contacts/people/1" ) ) );
```

This instructs Android to start the most appropriate Activity to dial the contact with ID number 1. You are unlikely to use this hard-code URI in practice, because contact number 1 is different from device to device. In order to find the right contact to dial, you need to query the Contact ContentProvider, using the syntax:

```
content://<authority>/<data_path>/<id>
```

Here is the data you need to include:

- **content://.** The standard, required prefix. This distinguishes the URI from other prefixes like http://. For ContentProviders, this will always be context://.

- **authority.** The provider name. Use short names only for Android default providers, like contacts. All third-party applications should use the fully qualified class name in lowercase.

```
context://com.recipes.basic.recipeprovider
```

- Authority is declared in AndroidManifest.xml with the **<provider>** tag and the **authorities** attribute:

```
<provider
android:name=".RecipeProvider"
android:authorities="com.recipes.basic.recipeprovider"
>
</provider>
```

- **data_path.** Identifies the kind of information requested, content://browser/bookmarks would specify that only bookmarks be included. This can include multiple segments, such as /recipe/cuisine and /recipe/category:

```
context://com.recipes.basic.recipeprovider/recipe/cuisine
```

- **id.** The unique identifier for a single record. This must correspond to a column named _ID in the data set. If absent, all records will return the following:

```
context://com.recipes.basic.recipeprovider/1
```

PITFALL

Not all ContentProviders are available without permission, which must be declared in the AndroidManifest.xml. In the case of Android core apps, these permissions are well documented in the SDK. In the case of other apps, you need to find the required permissions from the app developer's documentation or call getPathPermissions() for the provider while testing and debugging your app.

Requesting Data from a ContentProvider

In order to request information from a ContentProvider, you execute a query that returns a cursor object. Working with queries and cursors should be familiar from Chapter 17, and in the context of ContentProviders, the query methods are similar.

When possible, it is advantageous to import the Provider, as this allows quick access to its Constants. All of the imports necessary for this chapter follow here:

```
import android.content.ContentResolver;
import android.database.Cursor;
import android.net.Uri;
import android.provider.ContactsContract.Contacts;
import android.provider.ContactsContract.CommonDataKinds.Phone;
```

Now, you can write a query to return the lookup reference and display name of all the contacts on the device.

```
Uri contacts = Contacts.CONTENT_URI;
String[] fields = new String[] {Contacts.LOOKUP_KEY, Contacts.
    DISPLAY_
    ➡NAME};

Cursor cur = managedQuery(contacts,
fields,
null,
null,
Contacts.DISPLAY_NAME + " ASC");
```

Android provides two ways to execute a query on a ContentProvider, via the managedQuery() method on the Activity class and through the query() method on the ContentResolver class. In the case of managedQuery(), Android manages the entire life cycle of the cursor for you. Both methods take the same parameters and return the same results in a cursor object.

With the returned cursor, you can perform data binding to bring the results into a View or process them according to some other business logic. If the query has results, the cursor has stored them as a data table, with rows and columns. You can iterate this result set in a loop:

```
String displayName;
String lookupKey;
int nameColumn = cur.getColumnIndex( Contacts.DISPLAY_NAME );
int keyColumn = cur.getColumnIndex( Contacts.LOOKUP_KEY );
Uri aContact;
```

```
for( int i=0; i < cur.getCount(); i++ ) {
    cur.moveToPosition( i );
    displayName = cur.getString( nameColumn );
    lookupKey = cur.getString( keyColumn );
    if( "My Best Friend" == displayName )
    //send a Toast
}
```

To modify the data exposed by a ContentProvider, you must interact with the ContentResolver class, which provides the necessary methods to interact with a provider. You can instance a ContentResolver:

```
ContentResolver _ContentResolver = getContentResolver();
```

But this is not required. In most cases, you can take the results of a ContentResolver method into a URI property. To add a new phone number to the Contacts provider, you can call getContentResolver().insert() with the URI and ContentValues object to add. All ContentResolver methods take a URI as the first parameter.

```
ContentValues values = new ContentValues();
values.put( Phone.LOOKUP_KEY, lookupKey );
values.put( Phone.NUMBER, "555-123-4567" );
values.put( Phone.TYPE, Phone.TYPE_MOBILE );
Uri uri = getContentResolver().insert( Phone.CONTENT_URI, values );
```

 GOOGLE IT

The full list of ContentProviders supported by the Android SDK is available at http://developer.android.com/reference/android/provider/package-summary. html. This reference includes details on the permissions required, if any, and the format and type of data supplied by the Providers.

Working with the SDK supported ContentProviders can provide significant value to your app. This practice not only has the potential to improve your app, but it can also inform your ability to provide your own ContentProvider as you begin to understand how they are used by all Android developers.

Building Your Own Provider

As a feature, ContentProviders do not add a significant consumer "Wow" factor to your app; users are unlikely to notice until they begin seeing helpful links to their recipes in other apps that have implemented your service. For other developers, however, this feature is an eye-catcher.

If you want to incorporate weather, sport, or financial data into your app, you can save significant time and energy by using the better developed content from other apps. In this way, apps can improve other apps, and this attracts users and developers to your app and ContentProviders.

Prepare the Class

Your Provider class needs to extend the ContentProvider class, which has several required methods. If you remember the DBHelper class from Chapter 17, you might recall that you did not define a core Android class. Now you can. Your ContentProvider needs to access your database. Using your SQLite helper class as a component within the Provider can neatly integrate the codependent classes.

To complete the entire process, you need the following elements.

1. Add a <provider> tag to AndroidManifest.xml for .RecipesProvider.

```
<provider
android:name=".RecipeProvider"
android:authorities="com.recipes.basic.recipeprovider"
>
</provider>
```

2. Create a new RecipesProvider class or refactor DBHelper:

```
public class RecipesProvider extends ContentProvider { }
```

3. Allow Eclipse to add the missing imports and required methods, which include:

- **onCreate().** Called when the Provider is initiated. Returns a Boolean, which should be true if successful.

- **query().** Your implementation of the query() method can be called from managedQuery() or ContentResolver.query(). This must return a cursor with the dataset.

- **insert().** Called to insert new data into your Provider. Must return a URI for the newly inserted item.

- **update().** Called to update data within your Provider. Returns an integer with the number of rows affected.

- **delete().** Called to delete data. Returns an integer with the number of rows affected.

- **getType().** Called to request the MIME type of data at a URI.

4. Define a public static final URI for your ContentProvider. This provides greater readability and access to other developers:

```
public static final String AUTHORITY = "content://com.recipes.
basic.recipeprovider";
    public static final Uri CONTENT_URI = Uri.parse( "content://" +
AUTHORITY + "/" + _RecipesTable );
```

5. Define the column names that the Provider will return. For consistency, these should match the column names from your database. Remember, an _ID column is required.

```
public static final _ID = "_id";
    public static final RECIPE = "recipe";
public static final CUISINE = "cuisine";
public static final CREATED_DATE = "created_date";
```

6. Define MIME type constants. When specifying MIME types, the "vnd" qualifier is used instead of "com", as you would use for packages. A type should be specified for all matches vnd.android.cursor.dir as well as matches by unique ID vnd.android.cursor.item.

```
public static final int RECIPES = 1;
public static final int RECIPES_ID = 2;
public static final String CONTENT_TYPE = "vnd.android.cursor.
dir/vnd.recipes.basic";
    public static final String CONTENT_ITEM_TYPE = "vnd.android.
cursor.item/vnd.recipes.basic";
```

7. Define a UriMatcher to aid resolving getType() requests.

```
private static final UriMatcher _UriMatcher;
static{
    UriMatcher = new UriMatcher( UriMatcher.NO_MATCH );
    UriMatcher.addURI( AUTHORITY, "recipes", RECIPES );
    UriMatcher.addURI( AUTHORITY, "recipes/#", RECIPES_ID );
}
```

8. Implement the required methods and add optional ContentProvider methods as needed.

> **PITFALL**
>
> Methods such as insert() and update() can be called from any thread and should follow threadsafe guidelines. The onCreate() method is called from the main user interface thread and should execute quickly, to avoid ANR.

Eclipse supplies all of the required methods with their appropriate return types and a comment holder for you to begin supplying the Java code to complete them.

```
@Override
      public boolean onCreate() {
      // TODO Auto-generated method stub
      return false;
}
```

Implement the Methods

At this point, you have a perfectly formed ContentProvider. While it compiles and runs, it does not yet return any data from your app. To complete the process, each method needs to be fleshed out to perform the way end users and other developers expect.

If anyone interacts with your Provider, onCreate() must execute and is an excellent place to start. The RecipeProvider constructor already instances the database, so nothing else is required to happen when the Provider is created. In this case, onCreate() should simply return true.

While not necessary, you could check for a valid DBOpenHelper instance to validate your state:

```
@Override
public boolean onCreate() {
        if( null != _DBOpenHelper )
        return true;
        else
        return false;
}
```

After onCreate() completes, there are no other life cycle methods for the ContentProvider class. Beyond the internal use of the class for interacting with the database, other apps may call your Provider constantly or not at all. To err on the side of optimism, assume the former and implement the remaining methods.

Following onCreate(), getType() is the next, most straightforward method to implement:

```
@Override
public String getType( Uri uri ) {
        switch( uriMatcher.match( uri ) ){
                case RECIPES:
                return CONTENT_TYPE;
                case RECIPES_ID:
                return CONTENT_ITEM_TYPE;
                default:
                throw new IllegalArgumentException( "Unsupported URI: "
    + uri );
        }
}
```

As currently implemented, the RecipesProvider class responds only to URI requests for recipes or the specific ID of a recipe. You can expand this to include any number of filters, but for now getType() functions as a simple "all" or "one" qualifier on the Provider.

Ideally, other apps have queried your Provider for information before attempting to perform any other actions. The next, most practical method to implement is query(). Remember, other apps could call managedQuery(), which Android regulates, or they could call ContentResolver.query().

Here's how we make an app check with a Provider before it executes:

```
@Override
public Cursor query( Uri uri,
String[] projection,
String selection,
String[] selectionArgs,
String sortOrder) {
        Cursor result;

        if( _UriMatcher.match( uri ) == RECIPES_ID )
        result = getRecipe( uri.getPathSegments().get( 1 ) );
        else
        result = getAllRecipes();
```

continues

```
        result.setNotificationUri( getContext().getContentResolver(),
    ⇒uri );

        return result;
    }
```

As you have already implemented helper methods within the class, you need only call them inside query() to complete the method. As currently implemented, users can either receive a cursor with all recipes or a single recipe. The query() method and its corresponding helper methods need to adapt to user feedback.

As you might imagine, the remaining methods are small as you have finished all of the principle work in the last chapter. The insert() and update() methods take a ContentValues parameter. You can easily overload your helper methods to accommodate this.

Modify the current updateRecipes():

```
    public int updateRecipe(long rowId, String recipe, String cuisine,
    ⇒Date created )
    {
        ContentValues values = new ContentValues();
        values.put( _RecipeColumn, recipe );
        values.put( _CuisineColumn, cuisine );
    v   alues.put( _CreatedColumn, "created" );
        return db.update( _RecipesTable, values, _IdColumn + "=" +
    ⇒rowId, null );
    }
```

Add a new overload to handle updating a single recipe:

```
    public int updateRecipe( Uri uri, ContentValues values )
    {
        int count = 0;
        switch ( _UriMatcher.match( uri ) ){
            case RECIPES_ID:
                count = db.update( _RecipesTable, values, _IdColumn +
    ⇒"=" + uri.getPathSegments().get(1), null );
                break;
            case RECIPES:
                //Must implement an update for all recipes
        }
        return count;
    }
```

As you refine the logic within this class, you may want to design your overloaded methods to call each other in a hierarchy to ensure that all updateRecipe() calls follow identical paths of execution.

Your ContentProvider update() method is now clean and simple. Here's how it looks:

```
@Override
public int update( Uri uri, ContentValues values, String selection,
  ➡String[] selectionArgs ) {
      int count = updateRecipe( uri, values );
      getContext().getContentResolver().notifyChange(uri, null);
      return count;
}
```

You can just as easily perform the database update within this method, but as you build more code to interact with the database, it aids you to organize your methods in a way that keeps it easy to improve, maintain, and troubleshoot. By calling notifyChange(), any observers watching this Provider will be alerted that changes have posted.

Next, add an addRecipe() overload:

```
public long addRecipe( ContentValues values)
{
      return db.insert( _RecipesTable, null, values );
}
```

Modify the former addRecipe() to unify the execution path:

```
public long addRecipe(String recipe, String cuisine)
{
      ContentValues values = new ContentValues();
      values.put( _CreatedColumn, recipe );
      values.put( _CreatedColumn, cuisine );
      values.put( _CreatedColumn, System.currentTimeMillis() );
      return addRecipe( values );
}
```

You can now implement the ContentProvider insert() method as:

```
@Override
public Uri insert( Uri uri, ContentValues values ) {
      Uri u = null;
      long rowid = addRecipe( values );
```

continues

```
            if (rowid > 0 )
            {
                    u = ContentUris.withAppendedId( uri, rowid);
                    getContext().getContentResolver().notifyChange( u, null
    );
            }
        return u;
    }
```

The last required method to implement is delete(), but you might decide you want to postpone allowing other apps to delete your recipe data or that you want to write more rigid requirements for delete operations. In this case, you can leave Eclipse's automatically generated method in tact:

```
@Override
public int delete( Uri uri, String selection, String[] selectionArgs
    ) {
        // TODO Auto-generated method stub
        return 0;
}
```

You already created a deleteRecipe() helper method that your app has direct access to. You can easily incorporate this into the ContentProvider just as you did with insert() and update().

Provider Permissions and the Manifest

As you begin refining and tuning your app and its ContentProvider, you may desire to restrict certain actions, like write operations such as update() and delete(), based on permissions. If you plan to store sensitive information, you will want to add read permissions as well.

Permissions and other settings occur in the manifest under your <provider> tag. Review some of the following attributes which can customize the behavior, function-ality, and permissions of your app. All of the following attributes are optional and default to the value "true", unless otherwise specified.

- **android:readPermission.** A string value that indicates the permission which other apps must define in order to read or query the Provider. If undefined, all apps have read permission.

- **android:writePermission.** A string value that indicates the permission which other apps must define in order to write or insert/update/delete from the Provider. If undefined, all apps have write permission.

- **android:permission.** A string value that indicates both read and write permission. A defined readPermission or writePermission will take precedence over this permission.

- **android:enabled.** True if the system can instance the Provider, false if otherwise.

- **android:exported.** True if components of other apps may use this Provider, false if the Provider is available only to its own app.

- **android:initOrder.** If your app has multiple ContentProviders that have dependencies upon one another, you can specify the order in which they are loaded here using integer values.

- **android:label.** Specify a string resource as a friendly name for the kind of content provided. If empty, the label on the app is used by default.

- **android:multiprocess.** True if the Provider can run in multiple processes, false if the Provider is limited to a single process. "False" is the default.

- **android:process.** By default, all components of an app run in the same process, but each component can override this to specify a different process name in which to run.

- **android:grantUriPermissions.** True if conditions allow clients read or write permissions if the client has not requested permission from their AndroidManifest. If true, read or write permission is possible for the entire Provider. If false, permission exceptions can only be granted to specific URI paths added as additional attributes to the grantUriPermissions tag. "False" is the default value.

If you do define permissions of any type, be sure to provide documentation for other developers to make it easy for other apps to incorporate your Provider's content.

You now have a fully functioning ContentProvider that other developers can incorporate into their apps. With a well-developed Provider, you can position your app to serve or augment the central repository of recipes, inspiring quotations, historical photos, or anything else you can imagine to develop.

The Least You Need to Know

- For security, Android apps can only access their own, private data directory.
- Content between apps is shared through a ContentProvider.
- A ContentProvider and any required permissions must be defined in the AndroidManifest.xml.
- Providers return query content as cursors, just as database queries do.
- Design your Providers with thread-safety in mind.

Taking Your App to Market

Part

5

You came up with your app idea and worked hard to make it a reality: Now what? This part gets into the nitty-gritty of final debugging, serious testing, and, most importantly, introducing your app to the public via Google Market and other app stores.

Part
5

Taking Your
App to Market

You came up with your app idea and worked hard to make it a reality. Now what? This part gets into the nitty-gritty of final debugging, serious testing, and, most importantly, introducing your app to the public via Google Market and other app stores.

Comprehensive Debugging

Chapter
19

In This Chapter

- Logging messages with LogCat
- Debugging code in Eclipse
- Using the debug and DDMS perspectives
- Analyzing execution data with Traceview

If you have ever written a research paper, a company memo, or just an e-mail to a friend, you have probably thanked your spell-checker for catching the mistake with "thier," corrected to "their" and only later discovered you really intended "there." Just like your word processor, Eclipse catches your syntax and spelling errors. In the same way, just because your app compiles and runs, it does not mean you have not written a bug.

Unlike humans who can mentally "fix" language as they hear it, apps see a mistake and fail or worse—they do something completely unexpected. This chapter introduces the tools you need to output your app's actions to logs, and troubleshoot your app using the available debugging tools.

Logging

As a developer working with a new platform or language, you are likely to see a few or more fatal error screens. These are easy to do, from forgetting to define an Activity, Service, or ContentProvider in the manifest, to allowing a null reference to be passed, to an attempt to instance an object. These are also easy to track back to their source and fix by reviewing the output of the LogCat view in Eclipse.

In Eclipse, open the LogCat view from the menu by selecting **Window > Show View > Other > Android > LogCat.** When you run your app in the Emulator or on your device, the LogCat view quickly fills with system messages as the OS runs. In addition, you can see the entire *stack trace* of each error that the Emulator throws. See Figure 19.1.

Figure 19.1: *The classic Android fatal error pop-up.*

 DEFINITION

A **stack trace** lists each dependent method call in descending order from the last method called. In an error context, the stack trace begins with the error, followed by the method (A) which generated the error, followed by the method (B), which called method (A), and so on.

Fatal errors can usually be identified and fixed by viewing the stack trace in LogCat and looking back at the code. Some obvious omission might become evident. This is not always the case, of course, and sometimes you will be faced with less obvious errors; clicking a button does not launch an Intent, and no error is generated in LogCat.

In this case, it is useful to generate your own log data, which is also output to the LogCat view (see Figure 19.2), using the Log class. Logs are nearly like toasts to yourself; they are visible only to developers running the app from a debug perspective. The Log class is quite simple with only one real function—to output messages.

Figure 19.2: *The LogCat view.*

Each of these log methods is overloaded. The first takes two String parameters: a tag and a message. The second takes the tag and message parameters with an additional Throwable, which can be used in try..catch blocks.

Here are some logs that will send us error messages:

- **Log.e().** Logs Error messages.
- **Log.w().** Logs Warning messages.
- **Log.i().** Logs Information messages.
- **Log.d().** Logs Debug messages.
- **Log.v().** Logs Verbose messages.
- **Log.wtf().** Logs What a Terrible Failure messages or conditions that should never happen.

Adding log output to your app is straightforward. Add the import:

```
import android.util.Log;
```

The first parameter of a log message is the tag, which provides the context for the class where the log is generated. This should be a constant within each class.

```
private static final String TAG = "This Class";
```

You then request log output wherever you anticipate points of failure. Frequently, this is the first line inside a method:

```
private void showFireworks( int fireworkId ) {
```

continues

```
                              Log.d( TAG, "Tried to show firework with id = " +
     ➥fireworkId );
                    //...
        }
```

If you start your app in the emulator and watch the LogCat view, you will see a message with the time, verbosity, tag, and message content every time this method is called. The LogCat view has a Filter that allows you to sift through the messages looking for specific times, tags, and messages.

You can leave Error, Warning, and Information messages in your finished app, but remember that these consume space. While you could remove these before preparing your app to market, what if there is an easier way? Using a special debug tag in the AndroidManifest.xml, you can explicitly launch your app in Debug mode, which the next section covers in more depth.

By implementing a distinct logging class, you can leave all of your log messages, from error to verbose intact without any fear of impacting your users.

This is how you establish a list of logging errors outside of your app:

```
    import android.util.Log;

    public class LogClass {
        public static void d( String tag, String msg ) {
                if( Log.isLoggable( tag, Log.DEBUG ) ) {
                        Log.d( tag, msg );
                }
        }
    }
```

Extend this class to include all of the overloads and logging levels that you want to implement. You can then instance a LogClass object, which produces only log messages if your app runs on a developer environment in debuggable mode.

Debugging Your App

One of the golden rules of programming is that any sophisticated app contains at least one error. You inevitably write bugs and defects, some of which you will catch early before you release your app to market, and others that users will report.

As we discussed in Chapter 3, you can enable debugging in your app by defining the <application> tag android:debuggable as true in the AndroidManifest.xml:

```
<application
...
android:debuggable="true">
```

You now have the ability to run your app in Debug mode. The cheapest time to catch errors is during development, so you should run your app in debuggable mode at least once to work with your app like a user might, just to observe the LogCat and surface any obvious errors that you might have overlooked.

To run your app in Debug mode, select your project in Eclipse and then select **Run > Debug As > Android Application.**

The Debug Perspective

With your debuggable app running, switch to the Eclipse Debug Perspective from the menu by selecting **Window > Open Perspective > Other > Debug.** Running your app in debuggable mode differs from running your app directly on the Emulator or device in a few key ways.

First, the debugger is attached to your running code. This enables Eclipse to detect errors as they happen and *break*. This enables you to inspect the running state of your app right as a particular block of code is being executed.

Eclipse allows you to configure how and when the debugger will interrupt code execution from the menu **Window > Preferences > Java > Debug.**

Second, running in debuggable mode allows you to control code execution by setting *breakpoints*. Whenever the debugger breaks on a line, you have to option to *step* through the code. Set a breakpoint in your code by right-clicking the line number of the code and selecting **Toggle Breakpoint.**

DEFINITION

The debugger is said to **break** when it halts code execution. The debugger will break on errors or on **breakpoints** defined by you. In the case of breakpoints, the break occurs before the line at the breakpoint is executed.

A **step** is a movement from the current break in execution to the next line to execute, which can be a single line or whole blocks of code. Stepping instructs the debugger to execute only the code within the next step before breaking execution again.

Finally, the debugger allows you to inspect the state of the app and the operating system. You can inspect running threads, processes, resources, and variables as their value definitions change. Within the Debug Perspective a few views assist this process:

- The **Debug View** enables you to view the currently executing threads; that includes controls for resuming code from a break and stepping. See Figure 19.3.

- The **Variables View** enables you to inspect the properties defined within the class on the thread selected from the Debug View at any point of execution in the stack.

- The **Breakpoints View** provides quick access to all of your defined breakpoints.

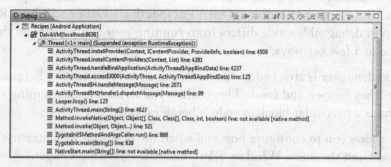

Figure 19.3: *The Eclipse Debug View.*

As you step through your code, you can follow the output of the LogCat view with the real-time state of your app.

Advanced Logging with Traceview

In the course of debugging your app, you may want more or improved logging detail that you can analyze on its own, apart from your debugging session, or that you can share with other developers to get feedback or insight. Android provides a special Debug class, which produces a special trace log.

Trace logs contain execution information from your app, which can be loaded with the Traceview utility to independently debug your app. To generate a trace, first import the Debug package in your LogClass:

```
import android.os.Debug;
```

The two relevant methods on this class are startMethodTracing(), which outputs a trace file to the virtual SD card in the emulator, and stopMethodTracing(). These certainly should not run in the market version of your app, and trace files provide far more information than is necessary for ordinary logging.

To keep them available and organized, add these convenient methods to your LogClass:

```
public static void startTrace( String tag, String fileName ) {
        if( Log.isLoggable( tag, Log.DEBUG ) ) {
                Debug.startMethodTracing( fileName );
        }
}
public static void stopTrace( String tag ) {
        if( Log.isLoggable( tag, Log.DEBUG ) ) {
                Debug.stopMethodTracing();
        }
}
```

PITFALL

Trace files output code execution information in great detail. Keep your trace files small and relevant by calling stopMethodTracing() as soon as you no longer need the additional trace information.

You will likely implement these helper methods only after you have performed some initial debugging and spotted a potentially problematic block of code. From there, you will go back and start trace logging inside the suspect blocks.

Once you have called your stopTrace() helper method, you can retrieve the trace file through the File Explorer view in the DDMS perspective, and open it with the Traceview utility (see Figure 19.4) located in your /android-sdk/tools directory.

Switching to the DDMS Perspective

Open the DDMS perspective by selecting **Window > Open Perspective > Other > DDMS.** Remember that perspectives in Eclipse are nothing more than convenient collections of views, which you can customize. The default views in the DDMS perspective are available only to Eclipse because the DDMS service is running part of the Android SDK and the ADT plugin, but this is the only real difference between DDMS and the Debug perspectives.

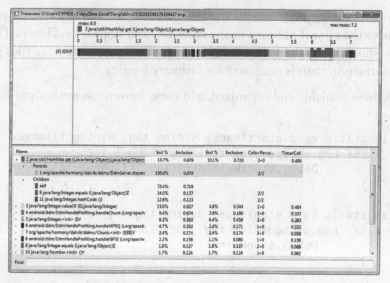

Figure 19.4: *The Traceview User Interface.*

 ANDROID DOES

Most of the views accessible through the DDMS view pass through the Android Debug Bridge (adb). While you do not need to use the adb directly if you use Eclipse, you should know that your DDMS actions correspond to adb commands. You can execute adb commands directly by launching /android-sdk-directory/tools/adb.

As discussed in Chapter 2, the Dalvik Debug Monitor Server (DDMS) perspective is designed to provide you direct access to more Android OS resources, and its views provide more specific kinds of information. The views are:

- The **Devices View** provides a list of all processes running on the Android OS, grouped by device—in case you have multiple Emulators or devices connected and running at the same time.

- The **Threads View** can be populated by selecting a process from the Devices View and clicking **Update Threads.** This lists all of the threads in the selected process with their statuses and running times.

- The **Allocation Tracker View** can be populated by selecting a process from the Devices View, then selecting **Start Tracking** and **Get Allocations.** This shows you the size of each instanced type according to which method or class instanced it.

- The **File Explorer View** provides you complete access to the Android file system. You can upload or download files to your package's directory, explore the system folder structure, or delete files.

You have already worked with the **Emulator Control View** to simulate GPS coordinates and incoming calls and text messages.

Defects in your code do not always generate errors that you would expect. Perhaps you implemented a ContentProvider that allows read/write access to other apps and you have an undiagnosed memory leak. The views in the DDMS perspective allow you to watch processor and memory utilization, which might be as important to debugging your app as stepping through code in the Debug perspective.

Debugging with Dev Tools

The last tool in your developer's debugging tool kit is the Dev Tools app, which is part of the Android SDK and installed by default in the emulator. The Dev Tools app displays diagnostic information overlaid onto your app as you run it. To get started, launch the emulator and select **Apps > Dev Tools > Development Settings.**

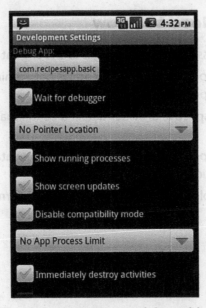

Figure 19.5: *The app development settings help you debug your program.*

From here, you can select the **Debug App** you wish to profile and define the information you want included on screen as you run your app. Here are the options, as shown in Figure 19.5:

- **Wait for debugger.** If selected, this stops the app you selected as the Debug App from launching until the Eclipse debugger has fully attached. This allows you to debug your app from the very first method called.

- **Show screen updates.** If selected, this flashes alternating pink and yellow colors to indicate that the screen is being drawn or redrawn.

- **Immediately destroy activities.** If selected, Android calls onDestroy() for each activity as it stops. Setting this allows you to quickly test your logic to save and restore state information.

- **Show CPU usage.** If selected, Android displays a CPU usage on screen.

You can also run the Dev Tools app on a physical device by copying the Development.apk from the emulator's /system/app/ directory and then installing the app to your device.

The Least You Need to Know

- Use the Log class to output messages to the LogCat view in Eclipse.
- Avoid sending log messages if your app is not running in debuggable mode by using smart helper methods.
- The Debug perspective provides detailed contextual information about code execution.
- The DDMS perspective provides comprehensive data about process and resource utilization.
- Use the Dev Tools app to get real-time system information while running your app.

Testing Your Apps

In This Chapter

- Reviewing the Android device diversity
- Developing a test plan
- Testing SDK versions in the emulator
- Understanding backward and forward compatibility

Every few years Microsoft or Apple releases a new version of its operating systems. The market suddenly divides between users of new computers that now ship with the new version of the OS and everyone else. The developer tension then arises: continue to provide backwards compatibility with the older OS or jump aboard the new OS with its latest and greatest features.

The pace of Android development is considerably faster. Android OS 1.0 was released in October 2008, and major OS updates are released every four to five months. With 200,000 Android devices sold every day, store shelves and consumer pockets are filled with many different devices running different versions of the OS. This chapter walks you through developing your app for the broadest possible market.

The Good News

Android developers used to fear the threat of Android fragmentation. In June 2010, approximately 25 percent of users had OS 1.5 (codenamed) Cupcake, 25 percent 1.6 Donut, and 50 percent 2.1 Eclair. By November 2010, that shifted to roughly 40 percent of users on Froyo, 40 percent on Eclair, and just 20 percent divided between Donut and Cupcake.

GOOGLE IT

Google releases the fragmentation or distribution of Android OS versions every month on the Android Developer site at http://developer.android.com/resources/dashboard/platform-versions.html. Periodically search for the current OS distribution to inform how you develop and test your apps.

This trend is encouraging for developers, because it indicates that hardware manufacturers are updating Android devices to newer versions of the OS. Maintaining app compatibility between Android OS versions has become significantly easier and is getting simpler with each quarter.

Your development prospects are promising. With almost no extra work, your Froyo targeted app is compatible with 40 percent of hundreds of millions of Android devices. With a little extra work, your Froyo and Eclair targeted app can reach 80 percent of the consumer market. How much more effort you want to spend to reach the last 20 percent is entirely up to you and depends on your needs and preferences as a developer.

Build a Testing Plan

Software development companies are likely to have separate departments for development, quality assurance, and unit testing, which probably include rigid criteria for executing test plans. As a developer working alone or in a small group, "testing" does not refer to that kind of formal process, though Android does have a testing API for this purpose.

Rather, testing means an informal process of building your app to target different configurations of Android hardware and OS versions, and then running your app on physical or emulated targets of those configurations. You should document what steps you take and issues you find, but this should be according to your own needs as a developer.

Before you design your tests, you will want to prepare a release strategy that answers these questions:

- Which Android OS versions will your app support?

- If your app uses new OS features, will you provide backward or fallback compatibility for older OSes?

- What hardware features will your app require?

- Which hardware features will you optionally support?

- How much of your testing will involve emulator targets?

- How many physical devices will you test?

By now, you have finished principle development on your app, and you have tested and debugged it on an emulator targeting Froyo. You can start drafting a simple test plan designed to answer this question: what core functions does your app need to perform to complete your goals for this version? In the case of your recipe app, your list might be this:

1. The app opens to the Main Menu.

2. The user can add a new recipe.

3. The user can see a list of recipes and edit an existing recipe.

4. The user can search for and find a recipe.

Presto! You have just defined your test plan. This may be all you require to complete version 1 of your app. As soon as you finish testing and submit your app to market, the development process begins anew for version 1.1. Now, you need to assemble your testing toolkit and commence testing.

Selecting Your Targets

Mobile phones are by far the most popular Android devices. Android tablets and Google TVs just started appearing on the consumer market in the fourth quarter of 2010. As a developer with limited time, focus most of your effort where it matters the most.

For now, it makes sense to target at least Android OS 2.1 and 2.2. Hardware is the next consideration, and screen size and density are arguably the most important features to test first. As of August 2010, Android devices are evenly split between medium and high density displays, ranging from 3½ to 4 inches in size.

Now, create some new emulator targets that fit within this range using some popular consumer phones as a reference point. The following table includes an OS 1.5 device so that you can see how your app might run. This does not need to be part of your test plan, but you can include it to remind you to evaluate the older OS later.

Emulators to Build for Test Run

AVD Name	OS	API/SDK	Display	Mimics Device
i1	1.5	3	HVGA	Motorola i1
nexus	2.1	7	QVGA	Google Nexus One
galaxys	2.1	7	WQVGA432	Samsung Galaxy S
droid2	2.2	8	WVGA800	Motorola Droid 2
evo	2.2	8	WVGA854	HTC Evo 4G

Create these emulator targets in Eclipse just as you did in Chapter 2, from the menu **Window > Android SDK and AVD Manager > Virtual Devices > New.**

ANDROID DOES

These AVDs can easily be reconfigured as you need them. Additionally, you can specify more options in the project's Run Configuration to alter network speed, latency, and other performance-altering options when the AVD is loaded in the emulator.

Working with Multiple Versions of Android

Before diving in and modifying your own app, take a moment to familiarize yourself with the process of changing Android OS targets and running them on different emulator AVDs. Following are the steps (see Figure 20.1):

1. Create a new Android project in Eclipse from an existing sample project, and choose Android 2.1 API 7 with LunarLander as the sample. The LunarLander project works on any version of the Android SDK, so it is a good choice for testing purposes.

2. Select the newly created LunarLander project and then select **Run > Run Configuration > Android Application > LunarLander.**

3. You will test against multiple AVDs, so make sure the AVD selection process is manual. Click the **Target** tab and set **Deployment Target Selection Mode** to **Manual.**

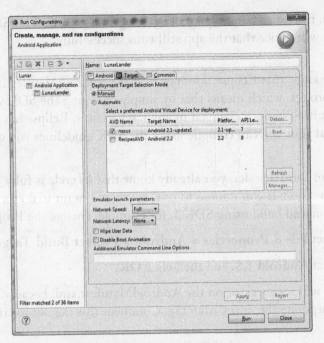

Figure 20.1: *Setting the SDK targets.*

4. Open LunarLander's AndroidManifest.xml and set the SDK targets to require SDK 7 but allow up to SDK 8:

```
<uses-sdk
android:targetSdkVersion="7"
android:minSdkVersion="7"
android:maxSdkVersion="8"
>
</uses-sdk>
```

5. Run the LunarLander project against your nexus (SDK 7) AVD and verify that the app runs successfully.

Keep in mind that the android:targetSdkVersion attribute is a guideline that is not enforced, but it should be between your min and max settings. The android:minSdkVersion and android:maxSdkVersion attributes are enforced. If you change android:minSdkVersion to "8" and attempt to run the project on your nexus AVD, you will see an error:

```
ERROR: Application requires API version 8. Device API version is 7
  ➥(Android 2.1-update1).
```

6. Run the LunarLander project again against your droid2 (SDK 8) AVD, and you will notice that the app still runs successfully.

You already know that Eclipse does not read the AndroidManifest.xml when it builds the project, which means that by simply changing the SDK version numbers, your project itself has not changed in any way—that is, Eclipse has made no attempt to verify that your app can compile using the SDK guidelines you define in the manifest.

In the case of LunarLander, you already know that its code is fully compatible with every SDK version. If you wanted to test, as you will want to do for your own app, that it does indeed build using SDK 3, for example, change the Project Build Target.

1. Select **File > Properties > Android > Project Build Target.**

2. Select **Android 1.5,** and then click **OK.**

3. You will see an error on the AndroidManifest.xml, because SDK 3 does not incorporate the <uses-sdk> tag. Comment this tag out for the moment.

The project compiles and runs on the emulator without an issue. Of course, as you already know from your experiences debugging, just because a project builds does not mean your app is free of bugs.

However, while the Android SDK changes frequently and sometimes dramatically, your Java code is likely to be more consistent. If your code compiles cleanly when using a different version of the SDK, it is a reason to be cautiously optimistic that there will be few compatibility issues.

Methods Change

As you have researched Android features, you have probably had the experience of scrolling through a list of methods on a class only to notice comments like, "This method is deprecated. Use method..." This indicates that the preferred way of operating has changed between versions of the SDK.

 GOOGLE IT

One of the nicest features about well-maintained open source projects like Android is that you can visit not only the entire source code of the operating system, but also its complete history. You can peruse the Android source at your leisure at http://source.android.com.

It is easy to use the preferred methods, constants, and classes as you develop forward, but if you need to develop compatibility backwards it means that the preferred reference you have just used in SDK 8 is not available in an older SDK. Obviously, if you want to target two SDK versions with your app, you want to use the preferred references relative to the OS version your app will run on.

The recommended way to accomplish this is to implement a wrapper class. The idea is that if you want to use a new class but are not sure whether it exists or not, you place the new class into a wrapper class that returns a trappable error if the class does not exist.

Suppose a fabulous new class is implemented in the next SDK version:

```
public class FabulousNewClass {

    private int _oldFirework;

    public FabulousNewClass( int oldFirework ) {
        oldFirework = oldFirework;
    }
    public int prepareFireworks( int newFirework ) {
        int f = _oldFirework;
        if( newFirework != f )
            f = newFirework;
        return( f );
    }

}
```

This class cannot be used on older versions of Android. In order to be able to write the code once and let it run on all devices, the class must be wrapped as follows:

```
class WrapFabulousNewClass {
    private FabulousNewClass _FabulousNewClass;
    static {
    try {
        Class.forName("FabulousNewClass");
    }
    catch (Exception ex) {
            throw new RuntimeException(ex);
    }
    }

    public WrapFabulousNewClass( int oldFirework ) {
        FabulousNewClass = new FabulousNewClass( oldFirework );
```

continues

```
}
    public static void init() {}
    public int prepareFireworks( int newFirework ) {
        return _FabulousNewClass.prepareFireworks( newFirework );
    }
}
```

This wrapper class simply attempts to instance the FabulousNewClass and throws an exception if the class does not exist. Now we can implement the FabulousNewClass through the wrapper:

```
public class FireworksDisplay {
    private static boolean _FabulousNewClassAvailable = false;
    static {
        try {
            WrapFabulousNewClass.init();
            FabulousNewClassAvailable = true;
        }
        catch( Throwable t ) { }
    }
    public void loadFireworks() {
        if( _FabulousNewClassAvailable ) {
            WrapFabulousNewClass.prepareFireworks( 1 );
            //New SDK is available, start the show
        }
        else
        {
            //Older SDK, must delay the party
        }
    }
}
```

In this way, you can continue developing with the newest Android functionality while allowing your app to run on older versions of the OS.

Be Mindful of the Future

You can be certain that for every hour you spend developing your app, Google has invested a hundred or a thousand fold more time developing the next version of Android. It is Google's ambition to constantly innovate and push the boundaries of the Android platform that makes Android development both exciting and intimidating.

Here are ways you can have the best experience debugging your programs:

- Follow the Android Developers Blog for updates and announcements.
- If you find an API in the source code that is not documented in the SDK, do not use it. It is a private API.
- Avoid overly complicated, deep-nested layouts with long hierarchies. Android's next performance optimizations might not appreciate them.
- Avoid all assumptions about hardware. Either require the hardware feature in the manifest or test for it.

On the flip side, that long array of "deprecated methods" is also a reason to be enthusiastic; with each new version of the OS, Android allows your existing code to still function while providing a gentle reminder that you need to update your references for the next version.

And More Tests

Testing and debugging often go hand-in-hand. Users may report a problem using a particular combination of hardware and OS that you may need to test and debug. As rigorously as you test your app before releasing to market, you will not catch every potential issue.

Android provides two additional tools to assist you with building and automating more comprehensive tests. First, the UI/Application Exerciser Monkey is a command-line tool that tests your app by randomly generating user and system level events. The Monkey runs on your emulator or device, producing a quasi-random but repeatable stream of clicks, touches, and gestures to mimic user interaction as well as interrupting your app with selected system events.

Second, Android provides a complete InstrumentationTestCase class to assist testing and instrumentation by creating test projects using the JUnit test methodology. A constantly evolving open source project, JUnit is less structured than Monkey, but fans argue that it provides more freedom.

GOOGLE IT

The JUnit framework is intended for self-described "Extreme Programming advocates". More information on the framework is available at http://www.junit. org. The Monkey tool is part of the Android SDK's tools package. Find complete documentation on these tools at http://developer.android.com/guide/ developing/tools.

With your testing complete, you have completed the last milestone of Android app development. Your next task is to publish your app and get it into the hands of eager users. The next chapter walks you through the publication options and the steps for submitting your app to the Android Market, the largest Android app store.

The Least You Need to Know

- Identify your test requirements and your version goals to build a test plan.
- Do market research to identify which Android OS versions you want your app to support.
- Test a variety of hardware and software configurations in the emulator.
- Provide seamless backwards compatibility with wrapper classes.
- Think about the future of Android OS development as you design your app.

App Markets and Beyond

21

In This Chapter

- Selecting an app storefront
- Preparing your app for the Android Market
- Final checklists
- Signing and submitting your app

The days of walking into a store and browsing through boxed software packages on shelves are all but gone. Most computer software can be bought and downloaded online in the same transaction. In the case of Android, even updates to the operating system are wireless-pushed to devices, requiring nothing but an Internet connection.

All of the apps available to install on Android devices are available only through markets, which you can shop in a browser or through dedicated market apps. This chapter introduces the Android Market, the largest Android app store, some alternative marketplaces, and the steps for submitting your app to market for download.

Available Android Markets

Of all the Android app markets available, Google's own Android Market is the oldest, largest, and most popular with over 100,000 apps available for download. Unlike the iPhone and iPad with their single iOS App Store, Google allows anyone to create their own storefronts and marketplaces for apps.

While you will likely want to focus on the Android Market first, you should familiarize yourself with some of the alternative stores.

- **Google.** The Android Market is the official Android app storefront hosted by Google. Visit http://www.android.com/market to browse popular apps and read the official developer policies.

- **Amazon.** Announced in the fall of 2010, Amazon will provide an Android app storefront. The full details of the developer program have not been announced, but this is arguably one of the most anticipated Market alternatives. Watch for more formal announcements in Spring 2011.

- **Verizon.** Verizon offers its own, curated V Cast Android storefront. The V Cast market is installed on most Verizon-sponsored Android phones. While much smaller than other markets, Verizon hopes to bring specific, high-quality, and well-reviewed apps to consumers. Visit http://developer.verizon.com/jsps/devCenters/Smart_Phone/index.jsp for developer guidelines and submission criteria.

- **Motorola.** The Shop4All market sponsored by Motorola targets Motorola-branded Android devices. Motorola offers additional development and testing tools as free Eclipse plugins, and it also offers early developer access to new devices. See http://developer.motorola.com/shop4apps for more information.

- **GetJar.** The second largest Android app storefront, GetJar allows developers to publish their apps and mobile optimized sites for free. They also provide tools for advertisers, which provide a draw to the market. Complete rules and instructions are available at http://developer.getjar.com.

- **Chrome.** Another market hosted by Google, this targets web apps. The Chrome app market is not yet available to the public, but developers are welcome to join and begin submitting their own web apps tailored to traditional and mobile browsers. Visit https://chrome.google.com/extensions/developer/dashboard for more detail.

GOOGLE IT

Many others have created Android app stores such as Andspot, SlideMe, and AndAppStore. If you want to attract a specific kind of user and audience, there is a good chance that you can find an app store tailored to your needs.

As with publishing articles, books, and music, app storefronts have differing policies on simultaneous submission, acceptance criteria, the types of payments you can collect from users, and what kinds of advertising are allowed. As you decide where to submit your apps, do the research to understand what effect this will have on your publishing rights.

Welcome to the Android Market

Google's marketplace is likely your first stop as a developer with a shiny new app. The Market provides a well-planned path to publishing your app, providing updates and collecting and responding to user feedback. Most of this chapter focuses on the specific requirements of the Market, but similar guidelines apply to the other storefronts.

Before you begin the process of registering for accounts and submitting your app, you will want to complete the app prepublication checklist.

- You have done a thorough test of your app.

- Your app has an icon and label defined in its AndroidManifest.xml.

- You have turned off debuggable mode and cleaned up your project's file system.

Registering Your Accounts

Signing up for an account and paying the developer's fee is the first step to publishing your app. You need a Google account to register for an Android Market Developer account.

1. Visit http://market.android.com/publish. If you do not have a Google account, you will be prompted to create one.

2. Google charges a small, one-time application fee ($25 U.S. as of Spring 2011). You will be prompted to register a credit card with Google Checkout to complete the payment.

ANDROID DOES

Developer accounts used to be free, but some less honest users abused the privilege. The fee is designed to be as inexpensive as possible while discouraging developers from "spamming" the Market with dummy accounts.

3. Once your payment has processed, you must accept the Android Market Developer Distribution Agreement, an *End User License Agreement (EULA)*. You should read the terms carefully on your own, but these are the key points:

    • There is no cost to distribute free apps

    • If you charge for your app, Google will keep 30 percent of the cost as a transaction fee and you will receive 70 percent. Apple currently has the same policy.

    • Users might receive a refund of the purchase of your app within 15 minutes of sale.

    • You agree not to distribute harmful or damaging software.

    • You alone are responsible for your app, updating and marketing your app and ensuring that it does not violate the copyrights of others.

    Overall, the agreement is unusually simple and fair. Unlike the Apple-branded App Store, Google does not review or reject apps, though it does patrol for malware, viruses, and adware.

4. If you want to charge for your apps, you need to set up a Merchant Account. The Merchant Account link is visible from your Developer home page. You need to provide credit card and bank information for Google to send you purchase payments.

Once you have completed the developer signup, you are ready to prepare your app for submission.

App Preparation

The app you have built and tested in Eclipse requires a few finishing touches before it is ready to submit to the Market. Review these final elements to prepare your Market-ready app.

• Add a EULA to your app (optional).

• Add licensing support to your app (optional).

• Add package and version identification (required).

- Create a cryptographic, signing key (required).

- Compile and sign your app (required).

- Use your signing key to register for a Maps API Key (required only if your app uses Google Maps API).

- Test your final app (highly recommended).

> **DEFINITION**
>
> An **End User License Agreement (EULA)** defines an agreement between you, the developer, and users who must agree to the EULA in order to use your app. A EULA can protect you as a developer by defining liability for misuse of your app or state licensing requirements for legal use of your app.

This checklist should serve as your guide every time you release a new app or an update to your app to Market.

1. **Making the EULA decision.** By registering to download Market apps, users have already agreed to some basic terms that protect both the developer and you. Sometimes, you may want an extra layer of agreement protection. With each update to your app, consider whether you should add or remove a EULA.

   For example, some users have tried to sue Google Maps because they followed navigation directions off cliffs or into oncoming traffic. A EULA is a protection buffer against the unintended uses of your app.

   Google even offers a EULA class that you can download and implement from http://goo.gl/tZEuA. You implement this in an Activity that launches the first time a user opens your app. The user then agrees or disagrees to the EULA. If the user disagrees, the EULA Activity continues to open, preventing access to the rest of your app, until the user accepts the agreement.

2. **Deciding whether to license your app.** Licensing is a mechanism to determine whether or not your app is authorized to run on a user's device at any given time. Google provides a free licensing service, that you can investigate at http://developer.android.com/guide/publishing/licensing.html.

3. **Name and Version your app.** Every Android Market app must have a unique package name defined in the manifest. This matches the name of your project's Java package by default. Once submitted, this name cannot be changed, so choose wisely.

All apps must have a version name and number that are specified using the android:versionCode and android:versionName attributes of your <manifest> tag in the AndroidManifest.xml. The versionCode attribute is most critical. No two versions of your app can share the same versionCode, and each update to your app must have a higher versionCode than the one before.

Version numbering is a matter of developer preference, as the numbers are only meaningful in how you define them. You might define a major new release of your app as 1.0, 2.0, and so on. These might occur every year. In between, you might have significant updates that add new features, which might be 1.1, 1.2, 2.1, and so on. You then might have maintenance or bug fix updates, which you might number 1.0.1, 1.2.5, and so forth.

Version numbering is entirely up to you, as long as the numbers continue upward in scale and do not repeat. The versionName, on the other hand, can be any description you like to describe the version, from "Version 1.0 (The Launch Release)" to "Minor bug fixes".

Certification

All Android apps must be signed with a digital encryption certificate before they can run on any Android device. You might pause and ask: then how does my app run on my device and in the emulator? Google provides the ADT plugin for Eclipse a standard, public certificate, which is used to sign all of your projects when you run them locally.

This public key does not work with devices that you do not use for Android development. In order to publish your app to Market, you need to create your own certificate and sign your app.

In Eclipse, this process could not be easier. Following are the steps:

1. Right-click your project, and select **Android Tools > Export Signed Application Package.**

2. Enter a location for the keystore and a password. See Figure 21.1.

PITFALL

It is critical that you keep your keystore and your password safe and secure. You need to use the same key and password for every update to your app. Do not lose them!

Figure 21.1: *Typing in your keystore and password for your Android app.*

3. Create the key. The Alias, Password, Validity, and any one other field are required. The validity represents how long the app will be active and up to date. To be on the safe side, Google recommends a validity of at least 25 years. See Figure 21.2.

Figure 21.2: *The app key that legally certifies the app.*

4. Select where to store your signed app. All Android apps are saved as APK files. This is the file you will upload to the Android Market. See Figure 21.3.

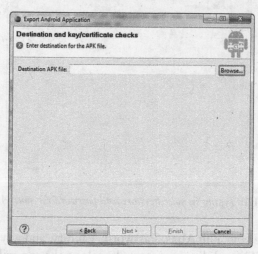

Figure 21.3: *Deciding where you will store the final app for the Android Market.*

You now have a valid certificate, good for this app and its updates for at least the next 25 years. If you also required a Google Maps API key, note that you need to go through this process twice. Once you have your certificate, you need to visit the Maps API key signup page at http://goo.gl/VeJMd.

With your API key, you will need to edit your MapView layouts to include your valid API, then sign and export the project again.

You now have everything in place to publish your app. It is highly recommended that you test this final package on a real device before publishing. The changes you have made between your final development testing and debugging are minor, but every change, no matter how small, is a candidate for a new bug.

And Release!

The marathon is nearly over and you are ready to cross the finish line! With your finished, signed, and testing Android app package in hand, log in to your Android Market Developer account and click the **Upload Application** icon.

The submission form requests icons, screenshots, images, video, descriptions, categorizations, a rating, and content information. You may want to organize all of the promotional content you wish to submit with your app by reviewing this submission list:

- Your app's APK
- App screenshots
- High-res app icons
- Promotional graphics
- Feature graphics
- Promotional video (YouTube link)
- Languages
- App title
- App description (325 characters)
- Changes with this version (325 characters)
- Promotional text (80 characters)
- App type
- App category
- Pricing
- Content Rating
- Locations
- Your website
- Your e-mail address
- Your phone number

PITFALL

All of this information is visible to the entire Market. Do not enter information, such as a personal phone number, that you do not want the entire Market to see.

One of the newest features in this form is the Content Rating, which is now required for all apps. There are currently four ratings:

- Everyone
- Low maturity
- Medium maturity
- High maturity

Any app without a Content Rating is classified as high maturity. Be sure to select a rating appropriate for your app.

As soon as you click agreement to the terms and click Publish, your app will be available in the Market. Of course, if your app is very large, this may take a few seconds to upload, but in Android Market terms, a completed upload equals publication.

Congratulations! You are now a published Android App developer!

Next Steps

You probably already have designs planned for the next big version of your app, but you also need to follow your app in the Market. The two most effective ways to improve your app standing—its rating and number of downloads—are to update frequently and connect with your users.

Frequent, meaningful updates improve your app's listing in the recently updated Market views. This increases your app's exposure and can attract new users. As long as your updates either introduce a new feature or address even a minor bug, you can steadily improve your standing.

One of the challenges of the new app Markets is anonymity. Unlike the days of shopping for software with a salesperson, downloading an app is an extremely impersonal affair for a user. When users comment on your app in the Market, it opens the door to real person-to-person communication—a scarce good.

React regularly and positively to constructive (and sometimes negative) comments. This will establish a rapport between you and your users, which can increase your app standing and provide valuable insight on ways to improve your app for your fans. Remember that no matter how excellent your app may be, nonconstructive negative comments happen to everyone—and you don't necessarily have to respond to them.

Advertising

The vast majority of mobile apps are free to download and use, but this does not mean that you, as a developer, cannot profit from the time that you have invested creating your app by making it freely available. Several advertising service tools are available for integrating ads in the mobile experience.

- Google AdSense
- Google AdMob
- Microsoft Ad Center
- Bidvertiser
- Mobclix
- Flurry

Google's own AdMob is a prime candidate for integrating advertising within your app, but the advertising space is diverse and growing rapidly.

So how does mobile advertising work in practice? Almost all of the providers define ways to delegate some portion of your app's screen space for the display of an advertisement. The syntax is usually provided for you to simply copy-and-paste into your app.

Whenever users run your app, if they have an Internet connection when viewing the screens you have delegated for ads, advertisements will automatically appear in that space. As a developer, you receive payment based on the number of ads your users click.

Both you and the advertisers have access to the click history. This enables you to alter the placement of the ads to try to generate more clicks, and it gives advertisers information that informs decisions about what kinds of ads to serve to users of different apps.

As a developer, you want to optimize your app experience for your users. Responding to feedback will be critical in determining how, when, and where or even if to place ads within your app.

The Next Generation

Another of the many golden rules of programming is that as soon as you finish your app, it is already obsolete. Rather than dismay you, this should inspire you. It means that new products and new features have inspired users to use apps in new ways, expecting bigger and better things from their devices. You can look at development as a constant challenge to keep up with or stay ahead of technological progress.

Most of the focus of this book has been placed on Android phones, but the operating system has bigger dreams. In November 2010, Samsung released one of the first mainstream Android tablets, the Galaxy Tab. Barnes & Noble released the latest Nook eReader, an Android device. Acer, Dell, HTC, and others have announced plans for Android-based tablets.

Google has been optimizing Android for tablets, most recently Android 3.0, code-named Honeycomb, which means new APIs and more consumer devices.

Furthermore, mobile phone technology is advancing rapidly. The next generations of devices will feature dual cameras, faster network connections, more sensors, and mobile payment technology. 3D displays will be here by the end of 2011.

Lastly, Google TV intends to put Android-capable devices inside of or attached to TVs in every living room. This brings great promise to the Android platform and its future.

Keep in Touch

As I have mentioned a few times throughout the book in passing, the entire source code for all of the content in this book is available for download, both by chapter as a zipped Eclipse project and as the complete source at http://goo.gl/GqGGr. If you have any issues or questions, submit them on the site. I will make every effort to answer them, and hopefully we can all learn from each other.

Thank you for reading, and may the wind be always at your back!

The Least You Need to Know

- Android apps may be published to any available storefront.
- Google's Android Market is the largest Android app store in the world.
- Market apps require a developer account which costs $25.
- All Market apps must have a signed certificate.
- Frequent updates and communication with fans are the key to success.

The Least You Need to Know

- Android apps may be published to any available store front.
- Google's Android Market is the largest Android app store in the world.
- Market apps require a developer account which costs $25.
- All Market apps must have a signed certificate.
- Frequent updates and communication with fans are the key to success.

Here you will find some of the common keywords and terms used throughout the text. This glossary is not comprehensive, but Appendix B will identify resources to find definitions for any terminology for which you need more information.

Android SDK   A collection of tools that integrate into your development environment enabling you to develop applications for any device (phone, tablet, or netbook) that runs the Android OS. SDK is short for Software Development Kit.

Application Programming Interface (API)   An API is a guide to writing code in a language. The API provides quick access to available elements of a language.

arrays   Systematic collections of objects, usually arranged serially in rows and columns.

asynchronous task   Task that executes on a background thread, separate from the main user interface thread of an app. The asynchronous thread needs to periodically update the main thread with the results of its progress.

Bluetooth   A wireless communication standard that allows data communication with any other Bluetooth-enabled device within range. Range is typically between 30 and 75 feet.

break and breakpoint   The debugger is said to break when it halts code execution. The debugger breaks on errors or on breakpoints defined by you. In the case of breakpoints, the break occurs before the line at the breakpoint is executed.

BroadcastReceiver   Listens for broadcasts sent by intents and can send notifications, run intents, start services, or update the display. BroadcastReceivers are started when a matching intent broadcast is sent.

class   A package that defines a blueprint for creating an object. The class contains rules for interacting with objects of its own class, and it contains the attributes and data available about each instance of object.

color In Android, a color is the hexadecimal combination of alpha, red, green, and blue. Alpha represents transparency. Opaque black would be ff000000, whereas transparent white would be 00ffffff.

Content Provider Makes data available to all apps that can be searched or modified. Android does not provide a common, central storage location, so providers are the only way to exchange information between apps.

cursor Contains an arbitrary number of objects that can be iterated like an array. Unlike an array, a cursor maintains a relationship with the query that generated it, enabling the cursor to request a query of its results and other operations.

Dalvik Debug Monitor Service (DDMS) Acts as a middle man between Eclipse and an Android device. Android allows for every process to be monitored in debug mode, and DDMS provides the infrastructure to both passively listen to Android apps as well as actively communicate with them.

data binding The process of joining two separate information sources together, frequently between a data source and XML or a user interface.

Drawable Any resource that can be drawn or rendered as a graphic on the screen. The Drawable class defines a few generic methods to interact with the object being drawn.

Eclipse The preferred Integrated Development Environment (IDE) for Android development. An IDE is a toolkit for writing, debugging, compiling, and deploying applications.

End User License Agreement (EULA) Defines an agreement between developers and users of an app. Users must agree to the EULA in order to use the app. A EULA defines waivers of liability for misuse of the app or state licensing requirements for legal use of the app.

Event Handlers Extend View components, such as Button or EditText widgets. Handlers provide method callbacks in the context of the parent widget.

Event Listener A View interface that contains a single callback method, such as onClick() or onKey(). These methods target the view and not components of the view.

Extensible Markup Language (XML) A web standard for storing data of any kind in a format easy to process programmatically. Almost all Android configuration is done with XML formatted documents.

Hello World   Often the first program developers write for new platforms or languages. The program is intended to demonstrate the simplest possible function: output a single line of text, "Hello World!"

Integrated Development Environment (IDE)   An application designed to assist your code-writing effort by helping compile your application, identify errors, and cross-reference functionality. IDEs help improve development efficiency.

interpolators   A class that lets you modify animation effects by increasing or decreasing the speed, bouncing, or repeating the effects.

Java Virtual Machine (JVM)   An application written for a specific platform that translates neutral Java code into code that the host OS can run.

life cycle   The time between when an object is created or started and when it is ended or destroyed.

Linux   A versatile, robust, and free operating system that is used to power a wide range of devices, from computers to TiVos and now mobile phones.

List View   Treated as a widget and a ViewGroup, a view that extends the ListAdapter class it uses to bridge data back and forth.

methods   A subroutine associated with a class or object that performs some action. They can return a value and optionally take parameters to customize their actions.

overloaded   When multiple methods exist with the same name when they actually require different parameters. Methods which serve similar purposes are sometimes overloaded to improve code readability and reduce code complexity.

pairing   When two Bluetooth devices establish a connection to each other. Once paired, the basic device information (name, address) is saved and communication with the device is possible.

permissions   The user giving an Android app the go-ahead to access certain features, such as location or the camera. Permissions are defined in the manifest, and the user is prompted to accept or deny an app's permission to the features on installation.

process   An instance of a program being executed, it contains the code to be executed and the activity. Processes in Android are composed of a single, primary thread by default.

receivers   Process information from transmitting stations, such as antennas, towers, and satellites. In mobile devices, this communication typically occurs over Radio Frequency (RF).

refactoring   The process of improving a source without changing its expected behavior. Reasons to refactor can include improving readability, decreasing complexity, or improving maintainability of code.

sensors   Return information collected from the device itself, such as touch screen input, rotation, and movement.

SharedPreferences   Simple information about the user stored as an XML resource. By default, this resource is private and can be accessed by only the calling activity.

Spec   A set of requirements for accomplishing a given task. They are as detailed and explicit as required for the person performing the task to complete the work and for peers to understand the goal.

SQL   The most widely used language for communicating with databases, and it is used to query, insert, update, and delete data. It also has the power to modify the structure of the database, modifying tables, columns, and permissions.

stack trace   In debugging, lists each dependent method call in descending order from the last method called. In an error context, it begins with the error, followed by the method (A) that generated the error, followed by the method (B) that called method (A), and so on.

step   In debugging, a movement from the current break in execution to the next line to execute. It can move forward by a single line or whole block of code and instructs the debugger to execute only the code within the step's increment before breaking execution again.

subclass   Inherits some of the methods and properties of a class. They share enough class characteristics to be considered related but are unique enough to deserve their own definition.

tab   A visual marker that allows multiple resources to be contained in a single screen with a visual cue to distinguish them.

thread   A unit of processing scheduled by an operating system to execute a series of tasks, usually in sequence. They are normally members of a process that represents a discrete activity or application.

toast   A view containing a message to the user.

tween   Animations defined in XML that describe transition effects to visually move objects from one state to another over some length of time.

Uniform Resource Identifier (URI)   An address for the location of some information. On the web, this could be a web address like http://www.google.com. In Android, this is likely to be a local address like content://contacts/people/1.

Universally Unique Identifier (UUID)   A random string that uniquely identifies public components of your app.

View (Class)   Responsible for rendering components of the user interface. It occupies a defined rectangular space and is responsible for drawing content and handling events in that space.

View (Eclipse)   Displays different types of information about your Workbench. Each has its own set of preferences and menu options and can be customized to your needs.

Resources, References, and Useful Websites

The web provides a vast number of sites dedicated to programming and assisting developers to connect and solve problems. While not comprehensive, this list provides some of the more popular sites for your reference. Remember, when in doubt, Google it.

Java and Object Oriented Language Resources

Many sites offer information for new developers to learn and refine their Java and Object Oriented Programming skills.

- **Oracle.** The official Java site contains many of the resources you need to get started in Java. http://download.oracle.com/javase/tutorial

- **W3Schools.** One of the largest developer sites available, W3Schools provides extensive guides and resources to standard technologies like XML, SQL, HTML5, and CSS. http://www.w3schools.com

- **Introduction to Programming Using Java.** A free, online textbook in its fifth edition, this book contains many useful guides to programming in Java. http://math.hws.edu/javanotes

- **Java Beginner.** Another free, online resource for Java programmers. http://www.javabeginner.com

- **The Object Oriented Programming Web.** Provides numerous links to free resources for object oriented programming and design, as well as references for Java. http://www.oopweb.com

- **iTunes™ University.** For iTunes users, the iTunes store provides a wealth of free lectures from prominent universities on every subject imaginable, including Java, Android, and programming concepts. http://www.apple.com/education/itunes-u

Google Services

Google provides extensive resources to document their products as well as tools to assist developing on Google's platforms.

- **Android Developer.** The official site of the Android SDK, where you can download all the current tools, browse the APIs, and read references and tutorials for developing in Android. http://developer.android.com

- **Official Android Blog.** The blog contains major announcements about new features, and changes to the Market and the SDK, but it also frequently provides in-depth guides to developing in Android. http://android-developers.blogspot.com

- **Google Code University.** Google provides a repository of lectures, classes, and tutorials for a variety of programming languages, concepts, and Android-specific resources. http://code.google.com/edu/languages/index.html

- **App Inventor.** A web based service that allows you to create and design Android apps in the browser and then customize and configure them on your device. http://appinventor.googlelabs.com/about

- **Android Cloud to Device Messaging Framework.** This service allows developers to push information from their servers out to apps. For more information or to sign up for access, visit http://code.google.com/android/c2dm

- **Google TV.** The next generation of Android devices will power the living room. Interested developers should visit the site to sign up for information and access to the necessary tools. http://www.google.com/tv/developers.html

- **Android Source Code.** The complete source code for the Android operating system. http://source.android.com

Android Developer Sites and Forums

The following sites dedicated to Android not only provide access to answers to your questions, but also keep you up-to-date with news, upcoming features, and devices that fuel your creativity.

- **Stack Overflow.** One of the fastest growing question-and-answer sites on the Internet, it has become a core resource in every developer's tool belt. To find Android specific questions, use the "android" tag. http://stackoverflow.com/questions/tagged/android

- **Android People.** Follows Android news and provides tutorials, sample code, videos, and a forum. http://www.androidpeople.com

- **Hello Android.** A popular Android developer site providing an app catalog, Android device information, and news. http://www.helloandroid.com

- **Android Academy.** A site dedicated to Android devices and news, it also contains helpful tutorials and sample code. http://www.androidacademy.com

Tools

The following tools will help new and experienced developers alike.

- **Eclipse.** The preferred IDE for Android development. http://www.eclipse.org

- **Eclipse Tutorials.** Guides for working with the Eclipse IDE. http://eclipsetutorial.sourceforge.net

- **Open Intent's Sensor Simulator.** Tool for simulating Android sensors using the emulator. http://code.google.com/p/openintents/wiki/SensorSimulator

- **3D OpenGL ES Libraries for Android.** Packages, APIs, and add-ons to work with complete 3D engines. http://www.openintents.org/en/libraries

- **Google Translate Java API.** A third party package that provides access to Google's Translation servers from Java. http://code.google.com/p/google-api-translate-java

- **The Complete Idiot's Guide to Android Development Source Code.** The complete source code for this book. http://code.google.com/a/eclipselabs.org/p/cig-android-development

Index

Numbers/Symbols

2D images. *See* drawables

3D images

  creating activities, 228-230

  environment resources, 230

  GL10 class, 227

  GLSurfaceView class, 226

  GLSurfaceView.Renderer class, 227

  modeling cubes

    coloring, 236

    creating, 230-233

    rotating, 233-236

  models, 230

  OpenGL, 225

  OpenGL ES, 226

  OpenGLView class, 228

  overview, 225

  polygons, 226

  primitives, 226

  triangles, 226

  vertices, 226

3G, 240, 248-250

4G, 240

/ (root folder), 22

A

About pages, 56-57

  adding activities, 58-59

  button, 61

  click event handling, 62-63

  layouts, 59-62

  manifest updates, 57

  text, 61

AbsoluteLayout, 37

accelerometer sensors, 215

access methods, 138-139

actions

  ACTION_CALL, 87

  ACTION_DIAL, 87

  ACTION_EDIT, 87

  ACTION_VIEW, 87

  call, 242

  intents, 87

  MotionEvent, 277

activities

  About pages, 56-57

    adding, 58-59

    button, 61

    click event handling, 62-63

    layouts, 59-62

    manifest updates, 57

    text, 61

  creating, 20

  defined, 32, 80

  executing in threads, 285

  gestures, 280

  life cycles, 32-33, 80-82

  OpenGL images, 228-230

  pausing, 82

  refactoring, 54-55

searchable
  building, 162-165
  defining, 161-162
Splash
  colors, 97
  creating, 94
  style, 98
  text, 97
  view, 95-96
switching between. *See* intents
transitions
  manually managing, 273-274
  methods, 272-273
  restart triggers, 270-271
user preferences, 71-74
view hierarchy, 94
adapters
  data binding, 311-312
  ListViewAdapter, 124
adb (Android Debug Bridge), 342
address links, 182
adLinks() method, 182
AdMob, 365
ADT plugin for Eclipse, 12-13
advertising, 365
alerts
  ContentProvider changes, 329
  dialogs, 170-173
  notifications, 297-299
Allocation Tracker view, 342
Amazon Android app storefront, 356
analog clock widgets, 176
Android Debug Bridge (adb), 342
Android Market, 355-356
  account registration, 357-358
  app names, 359
  app preparations, 358
  certification, 360-362
  content ratings, 364
  EULA, 359
  licensing, 359
  prepublication checklist, 357
  registration, 14

submissions, 363-364
version numbering, 360
website, 14
Android SDK (Software Development Kit)
  classes, 32
  defined, 4
  downloading, 11
  launching, 18
  locales supported, 262-263
  updates, 18
Android Virtual Devices. *See* AVDs
android.graphics package, 96
/AndroidManifest.xml file, 22
animateTo() method, 222
animations
  cubes, 233-236
  drawables, 106-108
  file types, 106
  frame, 106
  interpolators, 100
  listeners, 102
  nesting, 101
  pausing, 103
  resuming, 103
  splash screens. *See* splash screens
  tween, 99-100
ANR (Application Not Responding) dialog, 284
APIs (Application Programming Interfaces), 79
App Inventor for Android, 40
appendColumns() method, 315
appendWhere() method, 315
apps
  About pages, 56-57
    adding activities, 58-59
    button, 61
    click event handling, 62-63
    layouts, 59-62
    manifest updates, 57
    text, 61
  advertising, 365
  compatibility, 346

content ratings, 364
debugging, 25
   breakpoints, 27
   DDMS perspective, 28-29
   emulator, 27-28, 49-51
   running through the debugger, 51-52
Dev Tools, 343-344
fan communication, 364
Google Translate, 267-270
icons, 47-48
life cycles, 32-33
names, 19, 359
publishing
   certificates, 360-362
   developer account registration, 357-358
   EULA, 359
   licensing, 359
   names, 359
   preparations, 358
   prepublication checklist, 357
   submissions, 363-364
   version numbering, 360
quitting, 75
refactoring, 55
running, 25-27
styles, 65-67
updating, 364
version numbering, 360
arrays, 148
/assets file, 22
asynchronous tasks, 286-290
AsyncTask class, 286-290
audio
   embedding, 201-202
   recording, 202-203
AVDs (Android Virtual Devices)
   creating, 25-27
   running targets on multiple, 348-350

B

backwards compatibility, 350-352
BasicRecipes project. *See* Simply Recipes app
battery life, 212-213
bearingTo() method, 211
beginAnimation() method, 103
/bin file, 22
binding data, 311-315
bindService() method, 294
bitmaps, 106
Bluetooth, 240, 251-256
   classes, 252
   device discovery, 253-254
   enabling, 253
   pairing, 254
   permissions, 252
   querying paired devices, 253
   ranges, 252
   server connections, 254-256
   testing communication, 256
BluetoothAdapter class, 252
BluetoothClass class, 252
BluetoothDevice class, 252
BluetoothServerSocket class, 253
BluetoothSocket class, 252
breakpoints, 27, 339-340
breaks, 339
broadcast events, 89-90
BroadcastReceiver class, 89-90, 246
browsers
   advantages, 180
   converting text to links, 182-183
   embedding, 183-186
   history ContentProvider, 320
   opening websites, 180-181
   WebKit, 179
buildQueryString() method, 316
buildUnionQuery() method, 316
buttons
   About pages, 61
   adding, 37

click event handling, 62-63
quit, 75
radio buttons, 112, 115
rearranging, 40-42
toggle buttons, 113

C

cache directory access, 139
call actions, 242
CallLog ContentProvider, 320
CALL_PHONE permission, 241
CALL_PRIVILEGED permission, 241
camera
    control options, 195-197
    gallery, 200-201
    launching, 194
    permissions, 194
    preview, 193
    recording video, 198-200
    storing pictures, 197
    taking pictures, 194, 197
Camera class, 192
    control options, 195-197
    permissions, 194
cancel() method, 132
Canvas class, 108
capturing
    images, 197
    video, 198-200
carrier antennas, 206
centering menus, 40-42
certification, 360-362
check boxes, 73, 112
choosing
    development environments, 5-6
    platforms, 7
    targets, 19, 347-348
    themes, 65-67
Chrome app market, 356

classes, 32. *See also* specific classes
    backwards compatibility, 350-352
    defined, 78
    inheritance, 78
    multimedia, 192
    renaming, 54
    tabs, 127
    wrapper, 351
click event handling, 62-63
clock widgets, 176-177
close() method, 139, 308
coarse location services, 207
code assistance in Eclipse, 40
Color class, 96
colors
    cubes, 236
    defined, 96
    hex values, 97
    names, 97
    splash screens, 97
    values, 97
commit() method, 137
communication
    fans, 364
    features, 240
    network interfaces
        Bluetooth, 251-256
        Wi-Fi, 250-251
        Wi-Fi versus 3G, 248-250
    phones, 240
        call actions, 242
        call logic, 244
        call state changes, 244
        call state listeners, 245
        Dialer activity, 241
        emulating, 247
        MMS messages, 241
        native dialer, 241
        permissions, 241
        retrieving device/state information, 243
        text messaging, 241, 245-247

compatibility, 9, 346, 350-352
ConnectivityManager class, 248
contacts, 320
ContactsContract ContentProvider, 320
Content Ratings, 364
ContentProviders
  browser history, 320
  building, 324-330
  CallLog, 320
  change alerts, 329
  common, 320
  ContactsContract, 320
  data requests, 322-323
  defined, 32, 320
  deleting data, 330
  dialing contacts example, 320
  enabling, 331
  exporting, 331
  granting permissions, 331
  labels, 331
  load order, 331
  MediaStore, 320
  multiprocess, 331
  permissions, 321, 330-331
  qualifiers, 327
  queries, 327-328
  read and write permission, 331
  read permission, 330
  settings, 320
  state validation, 326
  syntax, 320-321
  udpating, 328-329
  UserDictionary, 320
  website listing of, 323
  write permission, 330
ContentResolver class, 323
context menus, 68
create() method, 173, 201
Cube class, 231
CubeRotate class, 233-236

cubes
  coloring, 236
  creating, 230-233
  rotating, 233-236
CubeSurfaceView class, 234
current location polling, 209
cursors, 307
customizing
  camera, 195-197
  layouts, 43-46
  search suggestions, 167

D

Dalvik Debug Monitor Service (DDMS), 29,
  341-343
data
  binding, 311-315
  storage
    external, 140
    internal, 138-140
    preferences, 136-137
databases, 301
  data binding, 311-315
  queries, 310-311
  SQLite. *See* SQLite
  tables
    adding columns, 303
    adding data, 304
    adding rows, 303
    creating, 302
    deleting, 303
    deleting data, 304
    field types, 303
    primary keys, 303
    queries, 303
    updating, 303
  testing/debugging, 316
date picker dialogs, 170
  activating, 174
  date value settings, 174

defining, 175
identifying, 174
launching, 175
native controls, 173
properties, 174
updating date values, 175
DBHelper class, 305-307
DDMS (Dalvik Debug Monitor Service), 29, 341-343
debugging
best practices, 353
breaks/breakpoints, 27, 339
databases, 316
DDMS perspective, 28-29, 341-343
Debug Perspective, 339-340
Dev Tools app, 343-344
emulator, 27-28, 49-51
enabling, 338-339
JUnit test methodology, 353
logs, 337-338
running through the debugger, 51-52
stack traces, 336
steps, 339
trace logs, 340-341
UI/Application Exerciser Monkey, 353
Debug Perspective, 339-340
Debug view, 340
delete() method, 330
deleteFile() method, 139
Density Independent Pixels (dp), 43
design
icons, 47-48
layouts
About pages, 59-62
buttons, 37, 40-42
common, 36-37
conditional qualifiers, 46
landscape layout definitions, 44
menus, 37-39
portrait/landscape display conditions, 43-45
resources file, 35

Simply Recipes Welcome screen, 34
widgets, 36
user interfaces
Fitts' Law, 119
ListView layout, 120-126
tabs, 127-130
toast messages, 131-132
visual utilities, 40
Dev Tools app, 343-344
developing for the future, 366
development environments
choosing, 5-6
tools
ADT plugin for Eclipse, 12-13
Android SDK, 11
Eclipse IDE, 10
JRE/JDK, 10
devices
AVDs, 25-27
Bluetooth, 253-254
compatibility, 9
displays
layout qualifiers, 46
portrait/landscape conditions, 43-45
features, 6-7
first, 8
orientation, 216
real-world simulation with emulator, 28-29
restrictions, 8-9
view, 342
dialing contacts, 320
Dialog class, 170
Dialog theme, 66
dialogs
alert, 170-173
Application Not Responding (ANR), 284
creating, 172
date picker, 173-175
defined, 170
dismissing, 172
life cycle, 171

native date and time controls, 173-174
 pre-configured, 170
 progress, 171
 removing, 172
 time pickers, 170
 viewing, 171
digital certificates, 360-362
digital clock widgets, 177
directories, 139-140
dismissDialog() method, 172
dispatching threads, 293
distanceTo() method, 211
divideMessage() method, 245
dock mode qualifiers, 46
doDialogDisplay() method, 172
doInBackground() method, 287
doTranslate() method, 269
dp (Density Independent Pixels), 43
dpi (screen pixel density), 46
drag and drop event handlers, 276
Drawable class, 106
drawables, 106
 changing views, 107
 lines, 109
 shapes, 107-108
 types, 106
DroidDraw, 40

E

Eclipse
 ADT plugin, 12-13
 code assistance, 40
 defined, 4
 download, 10
 emulator, 27
  debugging apps, 27-28
  launching, 27
  real-world simulation, 28-29
  running apps, 25-27

Foundation website, 17
Hello World, 24
 introduction, 16
 launching, 16
 layouts, 34
 perspectives, 17
 preferences, 22-24
Problems tab errors, 21
projects
 application names, 19
 building, 24
 creating activities, 20
 creating from existing example apps, 20
 debugging, 25
 files, 22
 min SDK version settings, 20
 names, 19
 package names, 20
 parameters, 19
 running, 25
 starting new, 19-20
 targets new, 19
 views, 20-21
resources, 17
views, 17
Welcome screen, 16
workspace, 17
Editor view, 21
EditText class, 112
e-mail address links, 182
embedding browsers, 183-186
emulator, 27
 AVDs, 348-350
 debugging apps, 27-28
 error page, 49-51
 launching, 27
 localization resources, 266
 location services limitations, 213-214
 multimedia support, 192
 phone calls, 247
 portrait versus landscape mode, 39

real-world simulation, 28-29
running apps, 25-27
Simply Recipes app, 39
targets, 347
enableCompass() method, 222
enableMyLocation() method, 222
enabling
ContentProviders, 331
debugging, 338-339
entire phase of activities, 80
errors. *See* debugging; troubleshooting
EULA (End User License Agreement), 359
events
broadcast, 89-90
handlers, 275
listeners
animations, 102
call state, 245
defined, 275
location, 208
text messages, 246
explicit intents, 87
exporting ContentProviders, 331
Extensible Markup Language (XML), 22
external storage, 140

F

fan communication, 364
field types, 303
File Explorer view, 343
fileList() method, 139
files
projects, 22
resources, 35
system storage, 138-140
filters, 88
findPointerIndex() method, 275
first Android device, 8
Fitts' Law, 119
foreground phase of activities, 81

forms
check boxes, 112
input definitions, 113
radio buttons, 112-115
spinners, 113-114
text fields, 112
toggle buttons, 113
fragmentation, 345
frame animation, 106
FrameLayout class, 36, 127
future of development, 366

G

/gen file, 22
German Zeit Online website, 267
GestureOverlayView class, 279
gestures, 274, 277-281
activities, 280
creating, 278
defining, 280
Gestures Builder API, 278
implementing, 279
Gestures Builder API, 278
getAccuracy() method, 211
getActiveNetworkInfo() method, 248
getAllNetworkInfo() method, 248
getAltitude() method, 211
getBackgroundDataSetting() method, 248
getBearing() method, 211
getBestProvider() method, 209
getBondedDevices() method, 253
getCacheDir() method, 139
getCallState() method, 243
getDataActivity() method, 243
getDataConnected() method, 243
getDefault() method, 245
getDefaultSensor() method, 215
getExternalFilesDir() method, 140
getFilesDir() method, 139
GetJar storefront, 356

getLastKnownLocation() method, 209
getLastNonConfigurationInstance() method, 272
getLatitude() method, 211
getLineNumber() method, 243
getLongitude() method, 211
getMenuInflator() method, 70
getNetworkInfo() method, 248
getNetworkOperator() method, 243
getNetworkType() method, 243
getOverlays() method, 222
getPathPermissions() method, 321
getPhoneType() method, 243
getPointerCount() method, 275
getPreferences() method, 137
getSensorList() method, 215
getSharedPreferences() method, 137
getSpeed() method, 211
getTime() method, 211
getType() method, 327
getVerticalFadingEdgeLength() method, 83
getVoicemailNumber() method, 243
getX() method, 275
getXPrecision() method, 276
getY() method, 276
getYPrecision() method, 276
GL10 class, 227
glClear() method, 227
glClearColor() method, 227
glClearDepthf() method, 227
glDisable() method, 227
glEnable() method, 227
glHint() method, 227
Global Positioning System (GPS), 206
GLSurfaceView class, 226
GLSurfaceView.Renderer class, 227
Google
    AdMob, 365
    Android Market. *See* Android Market
    EULA class, 359
    licensing service, 359

Maps
    access, 219
    adding content, 221
    API library access, 220
    controls, 221
    implementing, 220
    installing Google APIs, 219
    MapView class, 221
    permissions, 220
    Translate, 267-270
GPS (Global Positioning System), 206, 212
gyroscope sensors, 215

H

Handler class, 291
handling
    broadcast events, 89-90
    click events, 62-63
    drag and drop events, 276
    null values, 71
hardware
    communication, 240
    manifest declarations, 194
    receivers, 206
    requirements, 4
    sensors, 206-207
    targeting, 347
Hello World program, 24
hex values, 97
hierarchy
    ListView layout, 120-121
    views, 94

I

icons, 47-48
IDEs (Integrated Development Environments), 10

images
  3D. *See* 3D images
  capturing with camera, 194, 197
  file types, 106
  nine patch, 106
  public domain, 96
  storing, 197
  two-dimensional
    changing views, 107
    lines, 109
    shapes, 107-108
    types, 106
implementing
  ContentProvider methods, 326-330
  gestures, 279
  Google Maps, 220
  ListView layout, 123-126
  sensor services, 216-217
  service, 295
  Splash class, 101-102
  styles, 67
  voice searches, 166-167
  widgets, 176
implicit intents, 87
imports
  ListView layout, 124
  Splash class, 101
in (inches), 43
inches (in), 43
includeInGlobalSearch attribute, 160
info access, 139
inheritance, 78, 85
input
  check boxes, 112
  defining, 113
  Fitts' Law, 119
  gestures, 277-281
  multi-touch support
    actions, 277
    drag and drop handlers, 276
    MotionEvents, 275-277
  radio buttons, 112, 115

  spinners, 113-114
  storing as SharedPreferences, 148-151
  text fields, 112
  toggle buttons, 113
insert() method, 308, 329
InstrumentationTestCase class, 353
Integrated Development Environments
  (IDEs), 10
Intent class, 86-87
intentReceived() method, 165
intents, 86
  actions, 87
  broadcast, 89-90
  defined, 32, 86
  explicit, 87
  filters, 88
  implicit, 87
  location updates, 210
  requirements, 86
interfaces, 63
internal storage, 138-140
Internet
  Wi-Fi, 250-251
  Wi-Fi versus 3G, 248-250
interpolators, 100
isNetworkTypeValid() method, 249
isOpen() method, 308
isReadOnly() method, 308

J–K

Java
  defined, 77-78
  JVMs, 79-80
  Runtime Environment (JRE), 10
  SDK (JDK), 10
  tools, 10
JUnit test methodology, 353

keyboard availability qualifiers, 46

L

labeling ContentProviders, 331
LAN, 240
landscape layout-definitions, 44
languages
    alternate text translations, 263
    preparations, 264-265
    qualifiers, 46
    resource qualifiers, 264
    supported, 262-263
    testing, 266
    text translations, 267-270
last known locations, 209
launching
    Android SDK, 18
    browser, 180-181
    camera, 194
    date picker dialogs, 175
    Eclipse, 16
    emulator, 27
    searches, 165
layouts
    About pages, 59-62
    buttons, 37, 40-42
    common, 36-37
    conditional qualifiers, 46
    landscape layout definitions, 44
    ListView, 120
        hierarchy, 120-121
        implementing, 123-126
        inflation, 122
    menus, 37-39
    portrait/landscape display conditions,
        43-45
    scrollable
        blueprints, 115
        nesting ViewGroups, 116
        TableLayout, 116-119
    Simply Recipes Welcome screen, 34
        buttons, 37
        centering the menu, 40-42

common layouts, 36-37
conditional qualifiers, 46
landscape layout definitions, 44
menu, 37-39
portrait/landscape display conditions,
    43-45
resources file, 35
widgets, 36
SurfaceView, 193
views, 34
visual design utilities, 40
licensing apps, 359
life cycles
    activities, 80-82
    apps, 32-33
    dialogs, 171
    services, 294
light sensors, 215
Light theme, 66
LinearLayout, 36
Linkify class, 182-183
links, 182-183
Linux, 5
listeners
    animations, 102
    call state, 245
    defined, 275
    location, 208
    text messages, 246
lists
    arrays, 148
    ListView layout, 120
        hierarchy, 120-121
        implementing, 123-126
        inflation, 122
    preferences, 73
    ringtone selection, 73
ListView layout, 120
    hierarchy, 120-121
    implementing, 123-126
    inflation, 122
ListViewAdapter, 124

load ordering ContentProviders, 331
localization
    alternate text translations, 263
    defined, 261
    fatal errors, 264
    locales supported, 262-263
    preparations, 264-265
    resources, 262-264
    testing, 266
    text translations, 267-270
Location class methods, 211
location services, 205
    accuracy, 213
    battery life, 212-213
    coarse, 207
    emulator limitations, 213-214
    GPS availability, 212
    last location information, 209
    listeners, 208
    logical flow, 208
    maps. *See* maps
    methods, 211
    network provider declaration, 209
    polling current locations, 209
    privacy, 207
    receivers, 206
    requests, 208
    SDK support, 207
    sensors, 206-207, 216
    turning off updates, 212
    unavailable network providers, 211
    updating, 210-212
LocationListener class, 208-211
LocationManager class, 208-209
Log class, 337
LogCat view, 51, 336-338
logging
    adding, 337-338
    trace logs, 340-341
    viewing stack traces, 336
logic calls, 244
LunarLander project, 348

M

magnetic field sensors, 215
managedQuery() method, 322
managing threads, 290-291
manifest
    ContentProvider permissions, 330
    hardware/software declarations, 194
    updating, 57
maps
    adding content, 221
    API library access, 220
    controls, 221
    Google Maps access, 219
    implementing, 220
    installing Google APIs, 219
    MapView class, 221
    permissions, 220
MapView class, 221
market research, 7
markets
    advertising, 365
    Amazon, 356
    Android Market, 355-356
    app names, 359
    app preparations, 358
    certification, 360-362
    Chrome, 356
    Content Ratings, 364
    developer account registration, 357-358
    developing for the future, 366
    EULA, 359
    fan communication, 364
    GetJar, 356
    licensing, 359
    Motorola, 356
    prepublication checklist, 357
    submissions, 363-364
    updates, 364
    Verizon, 356
    version numbering, 360

media player service, 295-296
MediaPlayer class, 192, 201
MediaRecorder class, 192, 202
MediaStore class, 200-201
MediaStore ContentProvider, 320
menus, 68
    centering, 40-42
    context, 68
    options, 68-71
    Welcome, 37-39
Message class, 291
MessageQueue class, 291
methods. *See also* specific methods
    access, 138-139
    activity life cycle, 81-82
    AlertDialog class, 173
    AsyncTask class, 287
    backwards compatibility, 350-352
    ConnectivityManager, 248
    ContentProviders, 326-330
    defined, 78
    GL10 class, 227
    inheritance, 85
    Location class, 211
    log, 337
    MediaPlayer object, 201
    overloading, 210
    PhoneStateListener class, 244
    pointers, 275
    SmsManager class, 245
    SQLiteDatabase class, 308-310
    SQLiteOpenHelper class, 305
    SQLiteQueryBuilder class, 315
    TelephonyManager, 243
    toasts, 131
    transitions, 272-274
    View class, 83-85
microphone, 202-203
min SDK version settings, 20
mm (millimeter), 43
MMS (multimedia messaging), 240-241,
    245-247
MotionEvent class, 275-277

Motorola market, 356
multi-touch, 274
    actions, 277
    drag and drop handlers, 276
    MotionEvents, 275-277
multimedia
    camera. *See* camera
    classes, 192
    embedding audio/video, 201-202
    emulating, 192
    messaging, 240-241, 245-247
    microphone, 202-203
    permissions, 192
    photo gallery, 200-201
    SDK support, 191
multiprocess ContentProviders, 331
MyLocationOverlay class, 221

N

names
    applications, 19
    apps, 359
    AVDs, 26
    colors, 97
    icons, 48
    packages, 20
    projects, 19
native date and time dialogs, 173-174
navigation key availability, 46
NDK (Native Development Kit), 78
nesting
    animations, 101
    ViewGroups, 116
network interfaces
    Bluetooth, 251-256
    classes, 252
    device discovery, 253-254
    enabling, 253
    pairing, 254
    permissions, 252

querying paired devices, 253
ranges, 252
server connections, 254-256
testing communication, 256
Wi-Fi, 250-251
Wi-Fi versus 3G, 248-250
network provider declaration, 209
night mode qualifiers, 46
nine patch images, 106
notifications, 297-299
notifyChange() method, 329
null values, 71

O

object oriented programming (OOP), 78
onCallStateChanged() method, 244
onCellLocationChanged() method, 244
onClick() method, 63
OnClickListener interface, 63
onConfigChanged() method, 273
onCreate() method, 35
onCreateDialog() method, 171
    activities, 81
    ContentProviders, 326
    SQLiteOpenHelper class, 305
onDataActivity() method, 244
onDataConnectionStateChanged() method, 244
onDestroy() method, 82
onDismiss() method, 172
onDrawFrame() method, 227
onLocationChanged() method, 210
onMessageWaitingIndicatorChanged() method, 244
onNewIntent() method, 165
onOpen() method, 305
onOptionsItemSelected() method, 70
onPause() method, 82, 103, 226
onPostExecute(), 287
onPreExecute() method, 287

onPrepareDialog() method, 171
onProgressUpdate() method, 287
onRecieve() method, 90
onRestart() method, 81
onRestoreInstanceState() method, 137
onResume() method, 81, 103, 226
onRetainNonConfigurationInstance() method, 272
onSaveInstanceState() method, 137
onSearchRequested() method, 165
onSensorChanged() method, 218
onServiceStateChanged(), 244
onSignalStrengthsChanged() method, 244
onStart() method, 81
onStop() method, 82
onSurfaceChanged() method, 227
onSurfaceCreated() method, 227
onTabChanged() method, 128
onTouch() method, 275
onTouchEvent() method, 275
onTrackballEvent() method, 234
onUpgrade() method, 305
OOP (object oriented programming), 78
openFileInput() method, 139
openFileOutput() method, 138
OpenGL (Open Graphics Library), 225
    creating activities, 228-230
    ES (OpenGL for Embedded Systems), 226
    GL10 class, 227
    GLSurfaceView class, 226
    GLSurfaceView.Renderer class, 227
    OpenGLView class, 228
    polygons, 226
    primitives, 226
    triangles, 226
    vertices, 226
OpenIntents Sensor Simulator project, 218
openRawResource() method, 139
options menus, 68-71
orientation sensors, 216
OS
    distribution, 345
    targeting, 347

Outline view, 21
OvalView class, 108
overloading methods, 210

P

packages
    android.graphics, 96
    Explorer view, 21
    names, 20
Paint class, 108
pairing devices, 253
Path class, 108-109
pause() method, 201
pausing animations, 103
permissions
    Bluetooth, 252
    camera, 194
    ContentProviders, 321, 330-331
    defined, 192
    Google Maps, 220
    granting, 331
    location information, 207
    multimedia, 192
    system services, 248
    voice calls, 241
perspectives, 17
phone number links, 183
phones, 240
    emulating class, 247
    MMS messages, 241
    native dialer, 241
    permissions, 241
    text messages, 241, 245-247
    voice calls
        call actions, 242
        call logic, 244
        call state changes, 244-245
        Dialer activity, 241
        retrieving device/state information,
            243

PhoneStateListener class, 244
photos
    gallery, 200-201
    storing, 197
    taking, 197
pixels (px), 43
planning testing, 347
platforms, 7
playing audio/video, 201-202
pointers, 275
points (pt), 43
polygons, 226
portrait/landscape display conditions, 43-45
preconfigured dialogs, 170
predicting search terms, 167
preferences
    current view, 137
    Eclipse, 22-24
    editing, 137
    private, 136-137
    setting and retrieving, 137
    shared, 136-138
        retrieving, 142-148
        storing input, 148-151
    types, 73
    user, 71-74
prepare() method, 201
pressure sensors, 215
primary keys, 303
primitives, 226
privacy
    inactivity preferences, 137
    location services, 207
    preferences, 136
    searches, 159
Problems view, 21
PROCESS_OUTGOING_CALLS
    permission, 241
processes
    defined, 285
    executing activities, 285
    multiprocess ContentProviders, 331

prioritizing, 285-286
terminating, 285
programming
knowledge required, 4-5
OOP overview, 78
progress bars/dialogs, 171
AsyncTask, 287-289
self-managed threads, 291
projects
activities, 20
application names, 19
building, 24
creating from existing sample apps, 20
debugging, 25
files, 22
Hello World, 24
LunarLander, 348
min SDK version settings, 20
names, 19
package names, 20
parameters, 19
preferences, 22-24
refactoring, 54
running, 25
starting new, 19-20
targets, 19
views, 20-21
properties
date picker dialogs, 174
defined, 78
styles, 67
proximity sensors, 216
pt (points), 43
public domain images, 96
publishing apps
certification, 360-362
Content Ratings, 364
developer account registration, 357-358
EULA, 359
licensing, 359
names, 359
preparations, 358
prepublication checklist, 357

submissions, 363-364
version numbering, 360
publishProgress() method, 287
putInt() method, 137
putString() method, 137
px (pixels), 43

Q

qualifiers
ContentProviders, 327
languages, 264
queries
ContentProviders, 322-323, 327-328
databases, 303
SQL, 310-311
query() method, 310-311, 322
ContentProviders, 327-328
SQLiteDatabase class, 308
Quick Search Box, 157
quit buttons, 75
quitting apps, 75

R

R.java files, 35
radio buttons, 112, 115
Radio Frequency (RF), 206
RadioButton class, 112, 115
rating bar widgets, 178
rawQuery() method, 308
read() method, 139
READ_PHONE_STATE permission, 241
reading
access, 139
permissions, 330-331
sensors, 215-217
real-world simulation with emulator, 28-29
receivers, 206

reconnect() method, 199
recording
   audio, 202-203
   video, 198-200
refactoring, 54-55
region qualifiers, 46
RelativeLayout, 37
release strategies, 346
remove() method, 137
removeDialog() method, 172
removeUpdates() method, 212
renaming classes, 54
replace() method, 308
requestLocationUpdates() method, 209-210
requestRender() method, 234
requestRouteToHost() method, 249
requirements
   development environments, 5-6
   hardware, 4
   programming knowledge, 4-5
   tools, 10-13
/res file, 22
researching markets, 7
reset() method, 201
resources
   3D environments, 230
   Android developer website, 10
   Eclipse, 17
   files, 35
   icons, 47
   layout qualifiers, 46
   localization, 262
   searches, 167
   thread management, 290
responsiveness, 284
restart triggers, 270-271
restoring view preferences, 137
resuming animations, 103
retrieving preferences, 137, 142-148
   arrays, 148
   collections, 144
   declaring variables, 144
RF (Radio Frequency), 206

ringtone selection lists, 73
root folders, 22
Runnable class, 291
running apps, 25
   debugger, 51-52
   emulator, 25-27
   Simply Recipes, 39
runOnFirstFix() method, 222

S

Scale Independent Pixels (sp), 43
screen
   aspect qualifiers, 46
   orientation, 43-46
   pixel density (dpi), 46
   sizes, 46, 127
scrollable content
   blueprints, 115
   nesting ViewGroups, 116
   TableLayout, 116-119
SDK
   installing Google APIs, 219
   location services support, 207
   multimedia support, 191
Search Manager class, 158
   configuring, 159-161
   searchable activities, 161-165
searches
   activators, 165
   configuration control, 159-161
   custom suggestions, 167
   features, 158
   privacy, 159
   Quick Search Box, 157
   requests, 158
   resources, 167
   searchable activities
      building, 162-165
      defining, 161-162
   specifications, 159

suggestions, 160
    voice, 161, 166-167
searchSettingsDescription attribute, 160
searchSuggestAuthority attribute, 160
searchSuggestIntentAction attribute, 160
searchSuggestIntentData attribute, 160
searchSuggestThreshold attribute, 160
sendDataMessage() method, 245
sending
    MMS messages, 241
    text messages, 241, 245-247
sendMultipartTextMessage() method, 245
sendTextMessage() method, 245
SensorManager class, 215
sensors, 214
    data, 218
    device orientation, 216
    implementing, 216-217
    reading, 215-217
    test utility, 218
    types, 206-207, 215
services
    building, 294-297
    declaring, 294
    defined, 32
    functions, 293
    implementing, 295
    life cycle, 294
    starting/stopping, 296
setAnimationListener() method, 102
setAudioEncoder() method, 203
setAudioSource() method, 203
setBackgroundResource() method, 83
setButton() method, 173
setCamera() method, 198
setCancellable() method, 173
setClickable() method, 83
setContentDescription() method, 83
setContentView() method, 35
setDuration() method, 132
setFocusable() method, 83
setHapticFeedbackEnabled() method, 83
setId() method, 84

setKeepScreenOn() method, 84
setLatestEventInfo() method, 298
setLongClickable() method, 84
setMessage() method, 173
setNextFocusDownId() method, 84
setNextFocusLeftId() method, 84
setNextFocusRightId() method, 84
setNextFocusUpId() method, 84
setOutputFormat() method, 203
setPadding() method, 85
setRenderMode() method, 234
setSaveEnabled() method, 85
settings ContentProvider, 320
setTitle() method, 173
setVerticalFadingEdgeEnabled() method, 83
setView() method, 131
setZoom() method, 222
shapes, 106-109
SharedPreferences, 136-138
    retrieving, 142-148
        arrays, 148
        collections, 144
        declaring variables, 144
    storing input, 148-151
Shop4All market, 356
showDialog() method, 171
SimpleCursorAdapter class, 311-312
Simply Recipes app
    About page, 56-57
        adding activities, 58-59
        button, 61
        click event handling, 62-63
        layouts, 59-62
        MainMenu.java file, 63
        manifest updates, 57
        text, 61
    adding ingredients/recipes, 130-131
    alternate text translations, 263
    AVD, 25-27
    browser
        converting text to links, 182-183
        embedding, 183-186
        opening websites, 180-181

ContentProviders
  creating ContentProvider class, 324
  deleteRecipe() method, 330
  getType() method, 327
  insert() method, 329
  notifyChange() method, 329
  query() method, 327-328
  updateRecipes() method, 328-329
data binding
  adapter, 311-312
  template, 313
  view object association, 313-314
database interaction methods, 308-310
deleting recipes dialog, 172
files, 22
icon, 47-48
ingredients list, 121-126
input definitions, 113
laying out
  buttons, 37
  centering the menu, 40-42
  common layouts, 36-37
  conditional qualifiers, 46
  portrait/landscape display conditions, 43-45
  resources file, 35
  Welcome menu, 37-39
  Welcome screen, 34
  widgets, 36
listing user entered recipes, 142-148
options menu, 68-71
radio buttons, 115
recipe created date dialog, 174-175
Recipe List button, 151-154
recipe table, 302
RecipeActivity, 20
refactoring, 55
running, 39
saving recipes, 148-151
scrollable content
  blueprint, 115
  nesting ViewGroups, 116
  TableLayout, 116-119

searching
  building searchable activities, 162-165
  configuration control, 159
  defining searchable activities, 161
specification, 141
spinner control, 114
splash screen
  adding images, 96
  center image animation resource, 100
  colors, 97
  completed code, 103-105
  header and footer text effect, 99
  pausing, 103
  redirecting to Welcome screen, 102
  resuming, 103
  Splash Activity view, 95-96
  Splash class, 101-102
  style, 98
  text, 97
  transitions, 99-100
tables, 303-304
tabs
  creating, 127
  defined, 127
  focus, 128
  RecipeNew class, 129-130
  required elements, 128
toast message, 132
user preferences, 71-74
views, 21
sliding drawer widgets, 178
SMS (text messaging), 240
SmsManager class, 245
software manifest declarations, 194
SoundPool class, 192
source code downloads, 18
sp (Scale Independent Pixels), 43
specifications
  defined, 141
  searches, 159
  Simply Recipes app, 141
spinners, 113-114

Splash class, 101-102
splash screens
    activity, 94
    adding images, 96
    center image animation resource, 100
    colors, 97
    completed code, 103-105
    header and footer text effect, 99
    pausing, 103
    purpose, 94
    redirecting to Welcome screen, 102
    resuming, 103
    Splash class, 101-102
    style, 98
    text, 97
    transitions, 99-100
    view, 95-96
SQL (Structured Query Language), 302
    queries, 303
    tables, 302-304
SQLite, 304
    Browser, 316
    data binding, 311-315
    database interaction, 307-310
    download, 316
    helper class, 305-307
    limitations, 315
    Manager, 316
    queries, 310-311
    query helper class, 315
    SQLiteOpenHelper class, 305
SQLiteDatabase class, 308-310
SQLiteOpenHelper class, 305
SQLiteQueryBuilder class, 315
/src file, 22
stack traces, 336
start() method, 201
startDiscovery() method, 254
starting
    projects, 19-20
    services, 296
startMethodTracing() method, 341
startRecording() method, 199

startService() method, 294
startUsingNetworkFeature() method, 249
Status Bar, 297
steps, 339
stop() method, 201
stopMethodTracing() method, 341
stopping services, 296
stopRecording() method, 199
stopUsingNetworkFeature() method, 249
storing data
    external, 140
    internal, 138-140
    preferences, 136-138
    SharedPreferences class, 136
Structured Query Language. See SQL
styles
    creating, 67
    defined, 65
    implementing, 67
    properties, 67
    SplashStyle, 98
    themes, 65-67
subclasses, 78
submitting apps, 363-364
SurfaceView layout, 193
system
    service permissions, 248
    version qualifiers, 46

T

T-Mobile G1, 8
Tabhost class, 127
TableLayout class, 37, 116-119
TableRows class, 116
tables
    adding
        columns, 303
        data, 304
        rows, 303
    creating, 302
    deleting, 303

deleting data, 304
field types, 303
primary keys, 303
queries, 303
updating, 303
tabs
classes, 127
creating, 127
defined, 127
focus, 128
RecipeNew class example, 129-130
required elements, 128
TabWidget class, 127
takePicture() method, 197
targets
AVDs, 26
changing, 348-350
choosing, 19, 347-348
running on multiple versions, 348-350
tasks
asynchronous, 286-290
responsiveness, 284
specifications, 141
threads, 284
Tasks List view, 21
TelephonyManager class, 243
temperature sensors, 216
templates, 313
terminating processes, 285
testing
backwards compatibility, 350-352
best practices, 353
databases, 316
JUnit test methodology class, 353
plans, 347
release strategies, 346
targets
changing, 348-350
choosing, 347-348
running on multiple versions, 348-350
UI/Application Exerciser Monkey, 353
text
About pages, 61
alternate language translations, 263

converting to links, 182-183
fields, 73, 112
messaging, 240-241, 245-247
splash screens, 97
translating, 267-270
TextView subclass, 82
themes, 65-67
Thread class, 291
ThreadGroup class, 291
threads
asynchronous tasks, 286-290
defined, 284
dispatching, 293
dividing tasks into, 284
managing, 290-291
processes, 285-286
view, 342
time pickers, 170, 179
timer widgets, 177
toasts, 131-132
toggle buttons, 113
tools
ADT plugin for Eclipse, 12-13
Android developer website, 10
Android SDK, 11
Dev Tools app, 343-344
Eclipse IDE, 10
JRE/JDK, 10
sensor test utility, 218
Traceview, 340, 341
UI/Application Exerciser Monkey, 353
visual design, 40
touchscreen type qualifiers, 46
trace logs, 340-341
Traceview utility, 340-341
traditional arrays, 148
transitions, 99-100
activity life cycle methods, 272-273
manually managing, 273-274
restart triggers, 270-271
triangles, 226

troubleshooting. *See also* debugging
  apps
    debugging, 25-28
    emulator error page, 49-51
    running through the debugger, 51-52
  databases, 316
  Eclipse Problems tab errors, 21
  null values, 71
  unavailable network providers, 211
tween animations, 99-100

U

UI/Application Exerciser Monkey, 353
unavailable network providers, 211
units of measure, 43
Universally Unique Identifiers (UUIDs), 255
unlock() method, 198
update() method, 308, 328-329
updating
  Android Manifest, 57
  Android SDK, 18
  apps, 364
  ContentProviders, 328-329
  databases, 303
  locations, 210
URIs (Uniform Resource Identifiers), 87
user interfaces
  accessibility, 127
  Fitts' Law, 119
  forms. *See* forms
  layouts
    buttons, 37, 40-42
    common, 36-37
    conditional qualifiers, 46
    landscape layout definitions, 44
    menus, 37-39
    portrait/landscape display conditions, 43-45

ListView layout, 120
  hierarchy, 120-121
  implementing, 123-126
  inflation, 122
preferences, 71-74
scrollable content
  layout blueprint, 115
  nesting ViewGroups, 116
  TableLayout, 116-119
tabs, 127-129
toast messages, 131-132
widgets, 36
UserDictionary ContentProvider, 320
UUIDs (Universally Unique Identifiers), 255

V

Variables view, 340
vector shapes, 108-109
Verizon Android storefront, 356
version numbering, 360
vertices, 226
video
  embedding, 201-202
  gallery, 200-201
  recording, 198-200
VideoView class, 192-202
View class, 36, 82-85
ViewGroups, 82, 116
views
  camera input preview, 193
  data binding, 313-314
  DDMS perspective, 341-343
  Debug perspective, 339-340
  dialogs, 171
  Eclipse, 17
  Editor, 21
  GestureOverlayView, 279
  hierarchy, 94
  layouts. *See* layouts
  LogCat, 51, 336-338

Outline, 21
Package Explorer, 21
photos/videos, 200-201
preferences, 137
projects, 20-21
scrollable
blueprints, 115
nesting ViewGroups, 116
TableLayout, 116-119
Splash Activity, 95-96
Tasks List, 21
toasts, 131-132
View class, 82-85
widgets, 36
visible phase of activities, 80
visual design utilities, 40
voice
calls, 240
call actions, 242
call logic, 244
call state changes, 244-245
Dialer activity, 241
native dialer, 241
permissions, 241
retrieving device/state information, 243
searches, 161, 166-167
text messaging, 241, 245-247
voicePromptText attribute, 161
voiceSearchMode attribute, 161

W

web URL links, 183
WebKit, 179
websites
3D environment resources, 230
Android
developer, 10
Market, 14, 356
SDK, 11

Chrome app market, 356
ContentProviders list, 323
Eclipse
download, 10
Foundation, 17
German Zeit Online, 267
GetJar storefront, 356
Google
EULA class, 359
licensing service, 359
Maps licensing agreement, 219
hardware/software manifest declarations, 194
JRE/JDK downloads, 10
JUnit test framework, 354
Motorola market, 356
OS distribution, 346
search behavior resources, 167
Sensor Simulator project, 218
source code downloads, 18
SQLite, 316
style properties, 67
thread management, 290
Verizon Android storefront, 356
View class subclasses, 85
WebView widget, 183-186
Welcome screens, 34
buttons, 37, 40-42
common layouts, 36-37
menu, 37-39
resources file, 35
widgets, 36
Wi-Fi, 240
3G comparison, 248-250
WifiManager class, 250-251
receivers, 206
widgets, 36
analog clock, 176
check boxes, 112
defined, 170
digital clock, 177
implementing, 176

input definitions, 113
radio buttons, 112, 115
rating bar, 178
sliding drawer, 178
spinners, 113-114
text fields, 112
time picker, 179
timers, 177
toasts, 131-132
toggle buttons, 113
WebView, 183-186
wrapper classes, 351
write access, 138-139
write() method, 139
write permissions, 330-331

X-Y-Z

XML (Extensible Markup Language), 22
buttons, 37, 40-42
Welcome menu, 37-39